THE **PEPTIDE** PROTOCOLS

THE **PEPTIDE** PROTOCOLS

VOLUME 1

A Handbook for Practitioners

WILLIAM A. SEEDS MD

Published by
Spire Institute
Geneva, Ohio

ISBN 9780578624358 (print)

Book design and production by Happenstance Type-O-Rama
Typset in Adobe Caslon Pro and Bebas Pro

First Edition

TABLE OF CONTENTS

INTRODUCTION

As an orthopedic surgeon used to the straightforward surgical protocols that mend broken bones and address soft tissue injuries, I have simply been amazed at how a handful of peptides has changed the nature not only of recovery, but of performance. I've been working with peptides for almost 10 years—at first helping my athlete patients recover faster and then helping them return to the playing field stronger, more agile, and more capable. I've also been fortunate enough to create tremendous, measurable outcomes with non-surgery patients. Indeed, I've developed an entire supplementary practice in which I consult with other specialists and use peptides as adjutants that achieve faster, better medical results. For example, I have used them on a traumatic brain injury (TBI) patient who went from being non-responsive to walking and talking; on a teen girl whose kidney disease completely resolved; on an ALS (amyotrophic lateral sclerosis) patient who has since recovered motor skills; and on a young woman with chronic myelocytic leukemia (CML) who was able to lower her white count so she could avoid tyrosine kinase inhibitors (TKIs), which decrease infertility, and eventually conceive and deliver a boy. These are all remarkable cases, but they don't have to be. Peptides offer us nearly miraculous opportunities to change how we treat illness and disease; they also offer us life-changing tools and strategies for preventing disease in the first place.

This handbook, the first in a series, unpacks how disease—both chronic and acute—occurs at the level of the cell and how peptides can halt disease by enhancing cell functioning. It then introduces a number of essential peptides that address cell dysfunction and loss of efficiency and how to safely and strategically use these peptides to address the cell and intervene in common, debilitating conditions. However, what is perhaps most paradigm-shifting is that when you look through this powerful lens of the cellular level, you come to realize that peptides offer us a radically new way to define aging. Indeed, aging, as we've come to accept it, is simply cell cycle arrest. Peptides can target this arrest and the reasons for it. This is a gross simplification, to be sure, but there is great truth in the simplicity of the cell.

As the first in a series, this book serves as a general overview of an initial group of basic, though powerful, peptides; it covers their uses and their mechanisms for action within the cells and provides protocol examples. Although thousands of peptides are now in existence, with many more being developed every month, this handbook offers the first-ever comprehensive explanation of how and why peptides work. *The Peptide Protocols, Volume 1* is designed as an introductory guide; it provides the necessary background for any trained physician or professional healthcare practitioner who is interested in supplementing their practice to achieve more favorable outcomes. Some biohacking and self-improvement researchers around the world will also gain significant insight into the influence of peptide signaling on the cell.

All the information and advice pertaining to peptides is substantiated by significant peer-review studies (which are cited in a comprehensive bibliography at the back of this book). For continued education that focuses on the development of, maintenance of, and interaction with peptides, updated research and training can be found at Seeds.md.

This handbook accompanies workshops and training modules that I have created for the American Academy of Anti-Aging Medicine (A4M), along with several professional training programs that stem from my work with the International Peptide Society (IPS), a foundational organization of physicians and healthcare providers who are committed to best practices in the use of peptides. More recently, my work with peptide training and medical education has led to the establishment of the SSRP education program, which will take practitioners to the next level by offering continued, in-depth educational programming directed at cellular medicine.

This handbook serves as an introduction to what peptides are, how and why they work in the brain and body, and how they can be used judiciously to improve health and outcomes. It is meant to be used in conjunction with professional training and continued research in the fast-moving arena of peptides. You can use this handbook as a guide as you become familiar with various peptides, their mechanisms, modes of use, and interactions. You can also use this guide as a way in which to introduce peptides to your patients and inform them of their options and choices. Indeed, I am a strong believer in empowering our patients with knowledge and confidence; such patients always make more successful outcomes.

In Part 1 of the book, we take a close look at the cell cycle, cell behavior, and what happens when cells don't get what they need and begin to morph or make bad decisions. When cells get to this stage, they become *senescent*—and that's what we are really after with peptides. By interfering with or stopping cell senescence when necessary, peptides give us the opportunity to help our patients prevent disease, recover faster, harness aging, and improve overall health.

We have also introduced a new section that addresses COVID-19 in the chapter "Preparing the Immune System in the Age of Viruses, Bacteria, and Other Pathogens" to underscore the significant relevance of peptides, nutrition, and exercise in preparing people for a future in which viral and bacterial pandemics will increase. A key method for prevention and stemming the tide of disease is understanding the importance of cell efficiency and its role in optimizing immune modulation of the innate and adaptive immune responses.

In Part 2 of the book, we review the most commonly available peptides, their modes of action, their applications, and examples of protocols.

These topics will act as a guide that begins to unravel the puzzle of epigenetics and how we can utilize our current knowledge and understanding of how the genome reacts to epigenetic forces to create new phenotypes. Darwin was the first to recognize the process of evolution, which radically changed our understanding of the fate of the human race. Though Darwin had no knowledge of genes and the epigenetic forces that can alter phenotypes to change the programmed genomic makeup of a cell, he did understand that physical and behavioral adaptations could alter selective forces that would produce progeny, which in turn would predominate or desist in the environment, according to their level of fitness.

Almost one hundred years later, Richard Dawkins extended our understanding of this adaptable genome. In the 1970s, Dawkins recognized that the fundamental level at which natural selection acted was at the replicator (what he preferred to call *genes*) level—not at the species level. Since that time, scientists and physicians have assumed that the gene pool is the battleground on which phenotypic alterations competed for dominance.

Now it's time to update our understanding further and look at how and why this genome adapts, dies, or survives. Indeed, it's our responsibility to understand how this epigenetic signature change by phenotypic alteration is the basis for disease and aging. Ignoring these truths will have insidious effects in future generations. The cell and peptides lead us in this direction. And the time is now.

An Introduction to **Cellular Senescence**

1

It's Time to Redefine Aging

Since the beginning of time, humankind has aspired to take the reins on aging, defeat death, and discover—once and for all—a fountain that promises eternal youth. Our appetite for preserving youth continues and is perhaps even more intense at this moment than ever before. But often the approach to turning back the clock and holding onto youth is foiled. Hand and face creams, vitamin pouches, cosmetic procedures, and even growth hormones have never been able to deliver what they promise: an arrest of the decline in physiological and cognitive functioning that's associated with advancing age.

We are looking in the wrong direction.

Part of this confusion is based on our long-held adherence to the typical Western medical model that frames our understanding of biology in two basic ways: developmental and disease-oriented. This framing is rooted in the evolutionary lens: we are born, we develop to maturation so that we can reproduce, and then we die. Intellectually, we know that we have eclipsed this evolutionary schema—as evidenced by our long lifespans. Indeed, living way beyond our fertile periods is testament enough to the need to reframe our understanding of our own biology that is constrained by a view that the original evolutionary purpose is to design our genotypes and phenotypes.

The Western, allopathic model that has produced modern medical discoveries, procedures, inventions, drug therapies, and other treatment protocols is rooted in the study of disease, and its etiologies, complications, and risk factors. This means studying diseases themselves and trying to trace causes or triggers, assembling risk factors, analyzing complications, and hoping that a certain treatment method or protocol will erase the disease or arrest its progression. Most of the medical training that we all have received stems from this view of aging and disease, which inherently gives aging and disease the upper hand.

As functional and integrative physicians and healthcare practitioners, we have questioned this very premise for decades. Together, through our clinical experience and research, we have made great strides toward offering a comprehensive approach to redefining medicine, from one that is allopathic to one that is focused on preserving health and believing in the body's inherent capacity to heal based on its inherent drive for homeostasis. This drive toward homeostasis is just as strong—if not stronger at a cellular level—than any Darwinian mandate for survival of the species. Indeed, the power of the human cell, its very intelligence, is at the heart of understanding how and why we don't have to accept aging as our default. We don't have to passively wait for physical and cognitive deterioration and disease. We can harness what we know of our cell biology to empower our ability to adapt and live healthy, fruitful, fulfilling lives, regardless of our chronological number.

Over the last few decades, as more and more researchers and clinicians have investigated and begun to adopt a more integrative approach, they have also begun to reshape their research questions and approach to best practices. Why wait for disease? Why blast the body with chemo and radiation if cancer cells may be eradicated by the body's own immune system? What can we do to interrupt the epigenetic interactions that have spurred so many chronic diseases? How can we apply what we know through modern medical science to update our evolutionary design?

Some medical schools and training programs introduce basic tenets of preventative medicine, yet they still treat them as supplements or afterthoughts, thereby giving them only cursory attention. As both an orthopedic surgeon and integrative medical doctor, I use both traditions in my practice. But more and more, I am adopting a way to interact with my patients that puts prevention first. Of course, I am always motivated to improve my outcomes—I'm a surgeon after all. But I've come to realize—both through research and my clinical experience—that focusing on prevention offers many more robust, less harmful, nontoxic opportunities to not only prevent disease, but also to redefine the aging process as we have come to know it.

I have discovered a way for me, my patients, my family, and my medical colleagues to *embrace* aging. Specifically, I see aging as an opportunity to live our best lives. I feel fortunate to have arrived at this perspective because it's one that is empowering, optimistic, and accessible. In this book, I am going to introduce a novel way not only to think about aging, but to give you specific, accessible tools, strategies, and suggestions for protocols you can use to intervene with the conditions and diseases that are associated with aging.

And it's not that complicated. I've discovered a simple, yet revolutionary, approach to stimulate all of what prevention means—preserving health, achieving optimal homeostasis, off-setting the allostatic load, and supporting the integrity of the immune system. In short, we can avoid the downside of aging that triggers the onslaught of so many illnesses and disease conditions.

How would you like to understand more about how your patients can

- Improve and potentially reverse type 2 diabetes and metabolic syndrome?
- Prevent the precursors to heart disease from taking root?
- Avoid many cancers?
- Offset neurodegenerative decline?
- Minimize depression and anxiety disorders?
- Support the immune system?

- Fight against bacteria, viruses, and other sources of infection?

- Strengthen sexual functioning?

- Balance hormonal functioning?

- Preserve the vigor and strength of skin, hair, and nails?

- Protect cell efficiency and metabolic flexibility?

The key to this kingdom? Peptides.

Why peptides?

We know of at least 7,000 naturally occurring peptides in the body. Peptides are molecules that are a combination of two or more amino acids contained between an amine group and an H_2 group on one end, and a carboxyl group on the other. These amino acids are joined by what are called *peptide bonds*; peptides exist in all cells and are synthesized by the ribosome through translation of messenger RNA. The peptides are then transcribed into hormones and signaling agents. They're assembled and can become enzymes. They can also be ligands; they can be part of receptors. They can basically be any particular messaging part of a cell.

Typically, to be considered a *peptide*, a molecule must contain up to about 50 amino acids combined by peptide bonds. When a peptide contains between 50 and 100 amino acids, it is considered a *polypeptide*, and if more than 100 peptides are strung together to form a peptide molecule, it's typically called a *protein*.

All of these peptides have pharmacological profiles and intrinsic properties that offer selective messaging within any particular cell.

In the medical community, we've been utilizing peptides since the beginning of the 1920s. The first commercially available peptide in the United States was insulin, which is a sequence of 51 amino acids. It was first commercially available in 1923, making it the first peptide on the market. Insulin originally came from a glandular extract. In 1982, insulin made for human use became the first *recombinant* peptide that was made, meaning it was synthesized as a recombinant drug or a recombinant peptide.

In many ways, the discovery of insulin changed the world. We were not only able to start treating diabetes, we were also able to address and understand one of the most prolific metabolic diseases in the world. That changed medicine. It is no surprise that the researchers involved, Frederick Grant Banting and John James Rickard Macleod, won the Nobel Prize for the discovery of this peptide.

Since then, we have made huge advances in our understanding of what a peptide is. We know that they are naturally occurring molecules in the body and that a tremendous number of them are circulating in the body at all times. We know they play a crucial role as signaling agents within the cell cycle and that they assist in overall cellular functioning throughout the body. We have also begun to understand what happens when peptide production begins to ebb. But perhaps the most productive area of peptide research comes from the realization that, like insulin, we can re-create these naturally occurring signaling agents in the body. We also continue to look at other ways we may utilize these natural peptides to work the signaling system in a way that is advantageous to the cell. For instance, other peptides, like oxytocin, gonadotropin-releasing hormone, and vasopressin, have advanced our ability to preserve hormonal and cardiac health.

Presently, over 140 peptides are involved in therapeutic treatments and are being explored in clinical trials, and more than 500 therapeutic peptides are being used in preclinical development. We have over 60 U.S. Food and Drug Administration–approved peptide medicines on the market today—a number that continues to grow. Interestingly, back in 2011, the global market space for peptides was about 14 billion. As of 2018, it has expanded to well over 26 billion, and it is still growing.

The primary drivers for clinical research have been dominated by the tremendous and still-growing need to contain metabolic disease and oncology; these two areas are leading the research efforts in how to create peptides. However, lately the interest in peptides is extending beyond these two fields. Currently phase trials are going on in urology, pulmonology, pain, orthopedics, ophthalmology, infertility, hematology, gastroenterology, endocrinology, dermatology, and neurology, as well as in cardiovascular studies. There are antimicrobial and antiviral studies.

There are also studies being conducted on allergies, immunity, and bone and connective tissue. All together, these studies show that peptides present a burgeoning, interdisciplinary field of research. The findings are robust and are being used to bring about measurable, sustainable outcomes for diseases that up until now, we have been used to simply managing symptoms of.

These successful outcomes—some of which I have experienced in my own practice—are driving this interest in peptides. If we look back at what's been happening in the world of technology and drug development, historically the emphasis on research has been on discovering and creating molecules to be used in drugs. Scientists did not really move forward after the amazing discovery of insulin in the 1920s. Indeed, many physicians don't even realize that insulin is a peptide—a naturally occurring substrate in the body that was then re-created to help millions of people not die from type 1 diabetes. After insulin, we didn't really harness the power of peptides and what it means for medicine when a substance targets cells so specifically, has no toxicity, is recognizable and tolerated by the body, and causes no immune reactions.

Though many drugs and treatment protocols have helped to stem disease and reduce suffering, the fact is that they *all* come with side effects and secondary issues that can create problems 5, 10, even 15 years later. These medicines also cost millions of dollars to develop. They drive research funding. And some have been horrible blunders, such as the unforeseen outcomes of drugs such as thalidomide or some of the well-known issues with NSAIDs. The point I'm trying to make is that we've also seen, because of the costs in development, that the number of drugs developed has decreased considerably. You would think that with improvements in technology, biotechnology, and the combination of chemistry and computational drug designs, that we would be developing a lot more molecules for other treatment agents in medicine today.

We can spend time thinking about this missed opportunity . . . or we can empower ourselves and our patients to delve into and learn exactly how peptides work, and how they play crucial roles in human physiology—for instance, they are included in interactions between hormones, neurotransmitters, growth factors, ion channel control, ligands, and anti-infective properties of cell function. Perhaps the most appealing aspect of these naturally occurring peptides is how they function as therapeutics: when they are re-created, their specificity translates into excellent profiles for safety, tolerability, and efficacy in humans.

Some peptides are membrane permeable; some aren't. They are definitely, though, starting points for discovery, especially when we are currently looking at analogs of similarly structured peptides in the body to move into a new era of drug discovery. For example, there's been an amazing surge in the use of GLP-1 receptor agonists, which are peptides being used specifically to treat diabetes. GLP-1s are peptide chains (called *incretins*) that are made up of 37 amino acids; these are involved in insulin secretion and regulation and are produced in the small intestine. By harnessing this natural process and reproducing the peptides synthetically, we can treat the masses.

From a production standpoint, the biggest delay for this whole peptide revolution is the fact that peptides have a very, very short half-life when they're made or used in the body. They're signaled, they do their job, and they exit. As we've developed synthetic compounds, we've focused on creating substances that have a longer half-life to make the most of their impact and introduce them systemically. Increasingly, peptides, whether administered orally, subcutaneously, intranasally, or transdermally, have the power for deeper, longer-lasting effects.

We can do it, and we're getting better at it. In fact, we now have designs where we can make these peptides penetrate cells, the nucleus, and the mitochondria, and cross the blood-brain barrier. We now have mechanisms for altering the peptide, just a little bit—as long as it doesn't change its toxicity or potency—and giving it the ability to stay around a little longer and do what the cell wants it to do in its normal physiologic pattern.

Take, for example, a cancer cell. We can now take a peptide that has a specific pathway that it's going to find in a cell and use it to carry a specific chemotherapeutic or a particular molecule into the cell or its nucleus; it targets the cancer cell and no other cells. We have the ability to cross cell membranes. We can also use peptides in combination with medications to create adjutant therapies, with multiple routes of administration. That's the power of these peptides. Because the limits are seemingly endless, the literature on the use of peptides in treatment modalities is exploding all over the world today.

It's time to embrace this era in which we understand that maybe the body has it right and the cellular mechanism is perfectly capable of taking care of itself under every circumstance. It may need help sometimes to overcome some of the stressors that we've introduced from our environment or that are inside the body itself. But if we understand the mechanisms of these stressors on the cell, we can deal directly with the disease process of aging. The environment and landscape are quickly unfolding, and we can embrace these therapeutic modalities and options for acute and chronic conditions now. Peptides are being shown to be successful in aiding the regulation of blood and glucose, controlling insulin levels, and treating inflammatory diseases, brain diseases, cardiac disease, metabolic syndrome, weight issues, immune deficiencies, cancers, bone and joint problems, sleep disorders, anxiety, depression, and fatigue. We have multiple areas we can now address, not only in treating these chronic and acute conditions but also, potentially, in preventing them in the first place.

Most humans begin to cease making sufficient signaling agents by around age 30. This marks the cessation of development, a sign that our growing years are over. This is also when fertility begins to fall off. These are two biological vestiges of our evolutionary inheritance—when our lifespan was correlated more closely with birth, fertility, and death. Lifespans were shorter. Chronic diseases did not exist. And at the cellular level, there was less long-term use and therefore less need for us to continue to produce certain peptides associated with continued development, growth, and reproduction. After peak fertility, production falls off because, from an evolutionary point of view, it's deemed an unnecessary expenditure of energy. (Indeed, our bodies are inherently, instinctually energy efficient in their drive toward optimal homeostasis.)

However, just as we have not accepted the inevitability of an early mortality, have not forsworn wearing glasses for lack of good vision, or have not refused insulin when we are diabetic, we also need to question this seemingly inevitable cessation of peptide production.

In my practice, I have integrated peptides successfully to

- Offset cellular senescence.

- Hasten and ameliorate tissue, bone, and muscle healing, and function as a stand-alone therapy, in combination with physical therapy, or in combination with surgery and physical therapy.

- Support the immune system after injury and during repair.

- Regain neuromuscular functioning (ALS, MS, Parkinson's, and TBI).

- Restore kidney functioning.

- Minimize soft tissue, cardiac, kidney, liver, and pulmonary fibrosis.

- Preserve skin collagen.

- Treat anxiety and depression and improve cognition and memory.

- Improve recovery from training, changing the landscape for how athletes train.

So, what does all of this have to do with my new definition of aging? Aging as we know it is the number one factor in every disease we know. It itself is a disease—a disease of brain-body functioning marked by a loss of cellular efficiency. Peptides interfere with this negative spiral by giving cells what they need to continue to function and follow through with what they are preprogrammed to do: divide, grow, and mature. Peptides can make cells efficient again. Peptides can help direct cells to make good decisions. Peptides can allow the body's innate immune system to do its job of protecting against invaders and efficiently maintain health and homeostasis. And when all of this happens across the board within the body's cells, aging is just a number, not a recipe for disease.

2

The Intelligence of the Cell

In order to appreciate peptides, we must understand the simple yet profound intelligence of the cell, its cycle, its mechanisms for protecting itself and for making decisions, and its ultimate motivations. When we use the cell as our lens, we can uncover a roadmap that applies to all cells, systems, and processes. Indeed, all life begins with cells, which are preprogrammed to grow, replicate, and divide. If cells mutate or otherwise seem unlikely to survive, they set off their own internal "suicide" through the process of *apoptosis*. (You may recall that during apoptosis the cell is in the process of withdrawing from its environment and begins to disintegrate.) Also, part of the innate immune system is the process of *autophagy*, which occurs when the cells are signaled to come in and clean up any debris.

From this developmental perspective, a cell's primary objective is to be efficient in its goal of maintaining homeostasis. Indeed, regardless of where it is in the cell cycle, if the cell is allowed to do what it can do and it has the capability of utilizing all its mechanisms to react to any stressors in its environment, it will naturally formulate a plan so it can move forward. The cell needs stress to enter and proceed through its cycle. And although a cell necessarily encounters stressors every minute, every day, we know that its ability to adapt to stress varies based on exogenous and endogenous factors. For instance, as we age, we typically lose some of this capacity for adaptation—that's why stress is often linked to or explained as a cause of disease. It's not the stress itself; it's our capacity for managing that stress that weakens or decreases.

As we age, the cells of our brain and body slow down, they stop replicating, growing, proliferating—why? Because the peptides that signal this proliferation cycle begin to ebb. As a result, cells begin to either lose their efficiency or they begin making signaling mistakes. At the most basic level, this slowing down has a domino effect across the brain-body's systems, impacting all functioning, especially the immune system, which is always recruited when some kind of cellular change occurs or when homeostatic balance is disrupted.

Keep in mind how the immune system is intended to function: it is supposed to preserve homeostasis by combatting any internal (endogenous) or external (exogenous) stressors, toxins, or other agents that upset its capacity and efficiency for managing its allostatic load. If a cell loses some capability to handle the stress, it has difficulty making decisions or it begins to overcompensate and tries to make up for different environmental problems, for instance. Or if mitochondrial function is affected, cell signaling gets compromised. These are a couple of factors that can trigger illness and disease. For example, let's say someone has overeaten for years and is now on the edge of obesity with a BMI of 28; they don't exercise and otherwise live a sedentary lifestyle. This lifestyle creates enormous stress on the body: too much glucose is circulating in the bloodstream; the body is inflamed; the cholesterol is high; insulin resistance is being exhibited—all of these are the precursors of both diabetes and heart disease. Years of this situation make the cell inefficient and corrupt its signaling and decision-making acumen. The cell loses its metabolic flexibility.

Here's another example. Let's say an athlete is exercising, which typically enhances health, helps eliminate toxins from the body, strengthens the respiratory and cardiovascular systems, improves metabolism, and promotes neurogenesis and plasticity in the brain. All of this is good. At a cellular level, exercise is a form of stress that tests the cell's ability to utilize energy by up-taking glucose and producing ATP so cells can then utilize oxygen, and in turn provide that muscle tissue, brain cell, or organ with sufficient energy to function appropriately during exercise. However, when you overdo exercise and don't give your body enough time to recover and restore itself to homeostasis, that call for energy (required by exercise) taxes the cell and makes it less efficient. In this case, an outside stressor—exercise—has put too much load on the cell, undermining its efficiency. This leads to an imbalance in the innate and adaptive immune system, making it more TH2-dominant and increasing the possibility of upper respiratory infections, for example.

Basically, if we overwhelm a cell with stress, no matter what the stress is, it still has to respond and may have difficulty doing so, which has downstream effects. This is the same as what happens when we age: our cells' ability to manage stress (too much bad food, not enough deep sleep, a history of infections, etc.) compromises the cells' intelligence and immune response, which can lead to cells making bad decisions or mistakes.

This downstream effect can affect genetic expression as well. Genes can be vulnerable to negative changes, especially in regard to their phenotype. In response to environmental factors, genes may change their messaging and be upregulated or downregulated; this is the essence of epigenetics. When genes begin to change or adapt dynamically, cells begin to function differently, which means the phenotype of a cell begins to change, just as food or exercise can change the way those genes transcribe their proteins and enzymes.

Cells don't start developing problems in a vacuum. They develop problems or vulnerabilities based upon their capacity to deal with stress, whether it's bacterial, a viral stressor, environmental, specific damage or insult to the DNA, or problems with proteins in the cell that aren't being folded properly. Any of these stressors has the power to create intercellular or extracellular changes that disrupt cell functioning.

And when these insults attack the cell, the body relies on its built-in self-check system (autophagy and apoptosis) to stop itself, basically, from moving forward in this cell cycle process in order to take care of invaders and stressors. Essentially, the cell can check itself and say, "Okay, I'll hold off right now. I'll stop what I'm doing. I'll look inside and see if I have the capability of making myself better."

The liposomes, phagolysosomes, phagosomes, and peroxisomes are structures that can assist a cell's capacity to clean itself up. At the same time these structures are at work, the mitochondria are working hard to clean up or remove bad mitochondria. These processes—*autophagy* and *mitophagy*—are basically cleanup messaging systems that allow the cell to clean out debris that could be harmful to its environment.

A clean diet and regular exercise are a huge buffer to cell aging and dysfunction. Would you be surprised to know that only 5% of people adhere to a healthy regimen that maintains mitochondrial efficiency? Probably not. But as healthcare practitioners, that's what we are dealing with. Learning how to perceive and track how our patients are responding to stressors is an important feature in ultimately understanding what part of a person's cellular system is under fire, losing steam, or otherwise becoming overstressed.

However, there often comes a point—and this varies with each individual—when these varied processes of the self-check system can no longer keep up with the stressors. So, what happens when cells are postmitotic and have stopped dividing?

Cell senescence.

CELL SENESCENCE

Cell senescence was first discovered in the 1960s by Hayflick and Moorhead, who observed a number of cells that simply stopped dividing— they neither grew and divided nor progressed to cell death—these cells were arrested. This phenomenon changed our understanding of the cell cycle. Hayflick and Moorhead went on to define cell senescence as an indication of a cell's biological clock; they called it the *Hayflick limit*. At the time, the cell arrest of senescence was attributed to a progressive shortening of telomeres with each cell division. They understood this telomere erosion to be part of a physiological response to prevent genomic instability and DNA damage. Now we know that cell senescence is much more complicated.

First, senescence seems to have both positive and negative effects. It's been shown to play a positive role in embryogenesis and tissue remodeling, as well as in first-level immune response, helping the innate immune system with apoptosis and controlling potentially tumor-causing agents. These positive physiological characteristics, however, seem related to *transient* senescent cells—they go in, do their job, and exit.

Over the past several years, much more focus has been leveled at the negative, damaging, and inherently dangerous implications of *lingering* senescence. In essence, when cells turn senescent, they degrade cell signaling, create mitochondrial dysfunction, and set off a host of downstream negative effects that lead to disease. As Childs et al. point out in their 2017 study, senescent cells "disrupt normal tissue function by secreting factors that recruit inflammatory cells, remodel the extracellular matrix, trigger unwanted cell death, induce fibrosis, and inhibit stem cell function."

A general assumption is that senescence is a "natural" byproduct of aging. I am here to push back on that point of view: senescence is something we can go after, stop from happening, and sometimes even reverse. Again, aging as we have come to define it is not inevitable just as disease is not inevitable.

Let's take a look at the characteristics of senescent cells, the varied conditions that induce senescence, and the mechanisms that ensue. It's helpful to become familiar with characteristics that all senescent cells share, regardless of where they may be found. All senescent cells show

- Altered cell size and shape

- Accumulation of lipofuscin

- DNA damage to foci

- Loss of Lamin B1

- Upregulation of microRNAs

- Secretion of factors, including growth factors, cytokines, chemokines, and proteases. Together these factors make up what we now refer to as the *senescence-associated secretory phenotype (SASP)*.

- Senescence-associated changes in chromatin structure and function

- Age-associated changes in immune functions, noted as *immunosenescence*

Although we are making strides toward being able to identify senescent cells directly, there is no one biomarker. Instead, we rely on recognition of SASP and its covariants.

So how and why does senescence occur?

Senescence can be triggered by various mechanisms, through multiple pathways:

- Gradual erosion of telomeres, triggering a DNA damage response in the genome of the mitochondria

- Chronic physiological stress, including bad diet, lack of exercise, and stubborn insult to glucose metabolism. Downstream of senescence brought on by physiological stress, you will find oxidative stress, endoplasmic reticulum stress (resulting in unfolded protein response or UPR), mitochondrial stress (cytoplasmic chromatin fragments, or CCFs), and interferon-related responses.

- Acute injury or trauma to any area of the brain or body (anything from osteoarthritis in a knee to a TBI)

- As a result of anti-cancer treatments (radiation and chemotherapy) that purposely bring cell senescent onset; this is often referred to as *therapy-induced senescence*.

- Finally, senescence is the result of a systemic proinflammatory state, which has driven the immune system into hyper mode, degrading its ability to maintain homeostasis cellular protection.

Essentially, any of these senescence inducers are a form of stress on the cell and its systems. Yes, the cell requires stress to nudge it into its cell cycle. However, when any one or more of these stressors overwhelms the cell—in either their degree or their chronic-ness—the cell's ability to adapt becomes undermined. Cellular senescence is even more dangerous because it not only affects an individual cell, but it also affects nearby and distant tissue through chronic inflammatory response, reactive oxygen species, and insufficient apoptosis.

The bottom line is that the characteristics of the senescent cell—in particular its secretory phenotype—diminish its resistance to disease-causing stressors and cause stem and parenchymal cell dysfunction. (Indeed, stem cells are particularly vulnerable to damage from SASP because of the harmful effect on their microenvironment, or niche.)

The SASP is tricky, able to camouflage itself, and thereby become even more difficult to detect. When we understand the root causal mechanisms, however, we gain not only insight into how all of homeostasis and resistance to disease occur at the cellular level but also how we can use peptides to address the cell and its various environments that have been corrupted or made vulnerable by senescent cells. This approach—keeping our sights on the cell—is why, as an orthopedic surgeon, I remain able to work with so many types of patients and their physicians. It's also why I don't focus on the disease per se; I look at the cellular matrix for signs of senescence.

FOR SOME GOOD NEWS: There is a peptide, TA1, that can take away a senescent cell's camouflage.

HORMESIS

If cells stop dividing, then we can assume they need something that they are not getting. This is the basis of *hormesis*—what, in the cell environment, is causing it to go senescent? Is it lacking in proper nutrition? Is it not receiving signaling to continue to proliferate? Has it been overcome by toxins? We do know that cells can reach a *quiescent state*, where they stop proliferating; but here the cell still has the potential to enter the cell cycle again. Senescent cells are more or less irreversible because of a certain mechanism that gets triggered: *mTOR activation*. Although we need mTOR for basic anabolic processes, helping cells grow and proliferate, when mTOR gets activated by any one or more of the senescence inducers, it gets turned on too much, wreaking havoc on the MAPK pathways (ERK, JUN/JNK, and P38) that are in place to coordinate gene expression, mitosis, metabolism, motility, differentiation, and apoptosis.

What happens?

All sorts of cellular chaos. A cell continually influenced by mTOR pushes the cell to become hypertrophic, hyperactive, hyperfunctional, and signal resistant; together this causes profound phenotypical changes, making vulnerable areas more vulnerable, undermining the innate immune system, exacerbating inflammation, and ultimately causing disease states.

At a micro-level, a significant downstream effect of senescence on cellular efficiency and integrity is a dangerous loss of important nucleotide cofactor ratios (NAD+/NADH, NADP+/NADPH, Acetyl Co-A/Co-A, ADP/ATP), which are necessary for cell efficiency and metabolic flexibility. NAD+ (nicotinamide adenine dinucleotide) is a coenzyme in cells that functions as an electron acceptor, which is important to cellular redox. Triggered by a combination of decreased NAD+ biosynthesis and an increased consumption of NAD+, this loss of NAD+ and NADPH pooling pushes the cell to make bad decisions; in particular, it goes looking for NAD+ through the glycolysis pathway, a less efficient, more taxing process, and for NADPH through the pentose phosphate pathway. Further, this lack of sufficient NAD+ and NADPH downgrades our ability to produce antioxidants and reducing agents, which in turn increases a cell's oxidative stress, making it even less efficient.

Why are the NAD+/NADH and NADP+/NADPH ratios so important for cellular health? It has a very particular protective function for our SIRT gene, also known as our *longevity gene*. Essentially, downgrading NAD+ makes the SIRT gene vulnerable and less effective, which sets off a cascade of inflammatory responses that threaten the cell and its environment. SIRT genes activate the PGC1-alpha, utilize the AMPK pathway for fatty acid oxidation and mitochondrial biogenesis, and set off other transcription factors through a subtle feedback system that helps cells remain efficient. When cells lose their efficiency and metabolic flexibility due to poor substrate choice, they leak electrons through the electron transport chain and produce free radicals and reactive oxygen species, setting off all sorts of alarm bells and further inflammasome activation.

This is exactly what happens with type 2 diabetes. Type 2 diabetes is caused by chronic inflammation, hyperglycemia, and insulin resistance—all of which can lead to cell senescence and the associated secretory phenotype specific to the pancreatic beta cell. The SASP's cry for NAD+ and the resulting reliance on glycolysis creates chronic high circulating glucose, triggering insulin resistance and then type 2 diabetes. In this way, type 2 diabetes is both the cause of senescence and caused by senescence. In fact, most diseases have this reinforcing feedback relationship with senescence. Although we tend to think of treating type 2 diabetes through glucose control, this approach doesn't get at the essential inflammatory response at the root of the problem. However, when you drill down and set a treatment plan focused on the chronic inflammation, the proliferation of reaction oxidative species, and the mitochondria dysfunction, you are going after the SASP—the senescent cells that are spreading like wildfire. This strategy will have a much more lasting impact on type 2 diabetes, and may even reverse it.

Here's another example of the power of senescence to create disease. After a concussion, the brain perceives acute injury and the immune system responds by releasing cytokines, chemokines, and proteases to help the brain recover. But let's say that a young man has experienced numerous concussions during his adolescence and that some of these concussions have been noted but others have flown under the radar. After one more event, cytokines such as Interleukin 17 begin to proliferate in the brain, causing nitric oxide to increase and cross the blood-brain barrier. Cell receptors lose their signal sensitivity, exacerbating an already proinflammatory state. The brain perceives that it needs oxygen and so it upregulates the renin-angiotensin system (RAS). This causes the release of too much glutamate in the synaptic cleft of the neuronal synapse. The inflammatory RAS inactivation of the excitatory amino acid transport of glutamate out of the synaptic cleft into the astrocyte is needed for reprocessing back to glutamine so it can be reused in the presynaptic neuron. The result? Never-ending neuro-excitability that leads to cell death, which in this case, means neurodegenerative disease.

The good news is that peptides can intervene with senescence in one or more ways. When peptides are reintroduced, they give the cell a message to reset and correct any potential bad decision. Peptides also support the immune system and enable it to reinvigorate its capacity to fight off the proinflammatory state, infection, or a toxin assault. Peptides can also address the loss of NAD+ by promoting NAD+ biosynthesis and by preventing consumption by some of the greedy NAD+ culprits, including PARPs and CD38 Nadase. Peptides can assist the cell to regain its metabolic flexibility utilizing glucose and fatty acids (the Randle cycle), reversing insulin resistance, and mediating inflammation.

PEPTIDES

In general, we use peptides to

- Modulate appropriate inflammatory response
- Assist in cellular autophagy, mitophagy, and apoptosis
- Optimize mitochondrial function
- Reestablish or protect cellular efficiency
- Reestablish or maintain cell metabolic flexibility
- Maintain nucleotide cofactor ratios of NAD+/NADH, NADP+/NADPH, acetyl CoA/CoA, and ADP/ATP to maintain ultimate cellular redox
- Optimize P53, SIRT, and FOXO genes
- Monitor and influence timing of cellular senescence
- Reverse or inhibit epigenetic influences of the genome

Together these functions address cell senescence and ultimately prevent DNA mutations, which would lead to further senescence or cancer.

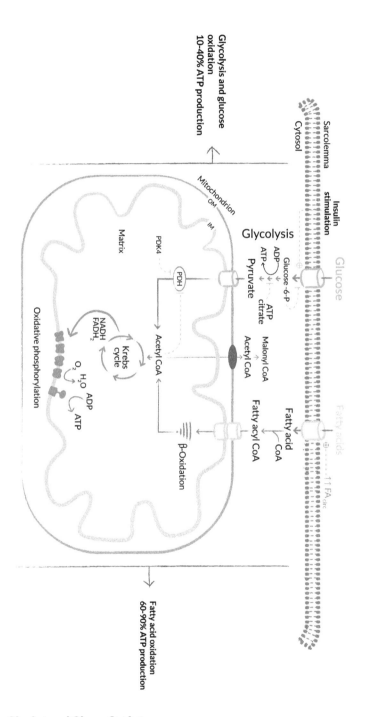

Glycolysis and Glucose Oxidation

MARKERS OF CELLULAR SENESCENCE

There is no one biomarker for cellular senescence. However, you can track the characteristics of the SASP, as well as a cellular pH of 6. One measurement tool called iTRAQ is developing different ways to recognize and quantify proteins, including peptides. Science is also headed in a direction of better understanding epigenetic influences on aging and disease. Technology has advanced with next-generation genomic sequencing, making it possible to better follow tens of millions of methylation markers that are related to specific diseases and aging. This technology allows for creating digital twins of ourselves and accurately predicting outcomes based on changes in nutrition, exercise, and therapies that may reverse epigenetic influences. I am of the belief that such technology will absolutely change how we accurately predict and monitor aging and disease with consistency and replicability in the future.

3

Delaying Senescence and Restoring Cellular Efficiency

One way to think about stopping senescence, halting its damaging spread, and otherwise preventing SASP from developing in the first place is to strategize about how to return the cell to homeostasis so that it can function as efficiently as possible. This is what is at the heart of cell optimization.

If we act on the concept that the cell needs to run efficiently to continue to provide the environment for the body to function optimally, then we gain the control to slow down the aging process, prevent disease, and ultimately build health. As we've seen in the previous chapters, loss of cellular efficiency leads to the progression of cellular senescence, which leads to proinflammatory states, causing progressive aging and metabolic issues that can further advance into disease processes that we are all too familiar with—osteoporosis, cardiomyopathy, cardiac disease, neurodegenerative disease, autoimmune disease (like diabetes), glaucoma, metabolic disease, and progression to cancer. The more we understand how a cell functions efficiently and what we can do to assist the cell in its ability to maintain this capacity to be efficient, the easier it is for the cell to *not* progress into that senescent state and create that SASP that then triggers the proinflammatory signaling agents; secretes cytokines, chemokines, proteases, and growth factors; and releases the extracellular matrix degradation proteins that lead to cellular inefficiency, senescence disease, cancer, and death.

Understanding cellular efficiency comes down to looking at the utilization of a substrate like glucose and maximizing the use of this substrate for the cell for cellular respiration, which entails utilizing oxygen and making ATP. If we look at the amount of oxygen consumed versus the amount of ATP produced, we can see what's called *mitochondrial coupling*. In this way, efficiency is a matter of consuming as little oxygen as possible relative to the amount of ATP that is produced, thus using the least amount of glucose. Efficiency also involves NAD+/NADH ratios and improving that pool because of the importance that NAD plays in supplying the hydrogen ions to that electron transport chain, enabling us to produce the hydrogen ion gradient and make ATP. This is the essence of homeostasis.

However, in order to do all of this efficiently, we also want to avoid creating increased reactive oxygen species or free radicals that can derail homeostasis. For this, we want to look for a more usable substrate than glucose for energy in a cell, and this leads us to fatty acid metabolism. By superseding the use of glucose and improving what we call *beta oxidation* or *fatty acid oxidation*, we not only improve the TCA cycle (aka the Krebs cycle) and mitochondrial respiration with oxidative phosphorylation, we also have the opportunity to produce a lot more ATP, which is converted to pyruvate and then to acetyl-CoA. Fatty acids can be transported across the mitochondrial membranes and further converted to acetyl-CoA. When we choose fatty acids as a substrate—the optimal substrate choice for mitochondria—we get an upregulation of beta oxidation.

In other words, the ultimate efficiency of the cell is really about leading the cell to choose the better substrate, short chain fatty acids. So how do we set the stage for the cell to choose the better substrate?

It starts with supporting the SIRT genes, which are known as our longevity genes. These SIRT genes can make proteins that are capable of influencing utilization of these substrates more efficiently. Part of that process entails an upregulation in the cytoplasm of PGC-1alpha that again points to the importance of maintaining an optimal NAD and NADH pool so that the available NAD+ can deacetylate.

PGC-1alpha sits in the cytoplasm and has to be deacetylated before it can be activated. It's held in the cytoplasm by a hypoxic-inducing factor one (HIF-1), which basically places a choke hold on the PGC-1alpha. It's not until the SIRT1 gene acts by releasing the hypoxic inducing factor (through the deacetylation of PGC-1alpha) that it can then transmit into the nucleus where it can be phosphorylated.

It's here inside the nucleus that its energy sensors are then activated, enabling the PGC-1alpha to phosphorylate AMPK and begin to transcribe other transcription messaging signals, such as the mitochondrial transcription factor (TFAM). This transcription factor undergoes translocation from the nucleus to the mitochondria, which is necessary to start mitochondrial biogenesis.

This transcription factor is also responsible for the mitochondria contributing its cytochromes to the electron transport chain to assist in ATP production. It's important to keep in mind that the nucleus and the mitochondria both contribute cytochromes within the electron transport chain and work together to produce ATP. As TFAM is upregulated, there is also an upregulation of PPAR gamma, delta, and alpha. Gamma, a main activator of the beta oxidation of fat, will also upregulate transcription factors to change the fiber type of muscle. In turn, this will improve the mitochondrial content of the fiber, thereby enhancing the efficiency of that muscle fiber in utilizing oxygen.

At the same time, there will also be an upregulation of FOX3 protein transcription factors, which also focus on longevity and antioxidant formation. This increase in other transcription factors provides an opportunity for decreasing phosphorylation in the nucleus of what is called Forkhead FOXO transcription factors, or just FOXO. When the FOXO proteins remain in the nucleus, they act like longevity genes and antioxidant genes and upregulate superoxide dismutase, catalase, and glutathione peroxidase, which means they improve the transcription of antioxidant enzymes and upregulate the nuclear factor—what's called *nuclear respiratory factor 1 and 2*.

All of these factors lead to improved mitochondrial function. By upregulating this metabolic response, the cell becomes more resistant to secondary assaults of oxidative species or antioxidants like H_2O_2, which can lead to DNA oxidative damage. Essentially, all of these factors increase and improve the efficiency of the mitochondria, enabling an upregulation of certain transcription factors in the nucleus that can then respond to an energy need or a stressor that the cell senses. The end result is cells that are more efficient at producing and utilizing energy, glucose, as well as ATP.

Again, starting with a SIRT1 gene, this upregulation helps the cell go through a series of steps, controlling a hypoxic inducing factor (HIF1) and releasing PGC-1alpha into the nucleus where it can then transcribe multiple transcription factors associated with cellular efficiency. These transcription factors (TFAM) improve mitochondrial biogenesis, increase cytochrome production for the electron transport chain, and increase the PPAR gamma, alpha, and beta production. As a result, they improve beta oxidation of fat, and thereby improve the optimal substrate, and assist in a phenotypical change of muscle fiber type; all of this leads to overall cellular and organ efficiency.

Keep in mind that if the cell starts losing its efficiency, as it does when it becomes senescent, and the electron transport chain is not working efficiently, free radicals or reactive oxygen species can be produced. This can lead to a mitochondrial injury or, potentially, to DNA damage that leads to further inflammatory processes. And that's really the key to the efficiency of the cell. If we can minimize any essential change of mitochondrial injury, we can improve the overall function of the cell, improve the ATP production, improve the utilization of the substrate, and make the cell the most efficient at utilizing oxygen. And, at the same time, we build the NAD to NADH pool, which is so important in cellular efficiency, DNA repair, and working with the longevity of SIRT1 genes.

Cellular efficiency has the ability to delay and even retard senescence in mitotic and post mitotic cells. But it also has the ability to regulate the reparative cells of the body—otherwise known as *stem cells*. When we refer to stem cells we are talking about the endogenous cells—hematopoietic stem cells (HSCs), mesenchymal stem cells (MSCs), pericytes—that are necessary for daily survival and that are called upon for tissue repair and regeneration. These cells and their lineage are regulated by the SIRT1 gene. Stem cell senescence is inversely related to SIRT1 gene activation. With stem cell senescence, reactive oxygen species (ROS) are high and the SIRT1 production is low, and the p53 protein (a tumor suppressor) is upregulated but inactivated because of low SIRT1 (which cannot be deacetylated). This low SIRT1 gene transcription under senescence also decreases FOXO1 transcription, which is an important regulator of the cellular antioxidant response (i.e., part of its innate defense system).

However, FOXO1 needs to be deacetylated to translocate to the nucleus where it can then increase antioxidant transcription of superoxide dismutase and catalase. Also, with the loss of deacetylation from low SIRT1, the powerful NRF2 translocation to the nucleus is also dampened, and this results in a reduction in the activation of the promotor region of the antioxidant-related response with a decreased transcription of Heme oxidase 1, NADPH quinone oxidoreductase1 (NQO1), catalase, and superoxide dismutase. In summary, stem cells rely on FOXO1 and NRF2 under the control of SIRT1 regulation to maintain control of ROS and stop cellular senescence.

A tremendous amount of research has been done to validate upregulation of the SIRT1 gene with allosteric activation by the polyphenol resveratrol. However, the human trials have failed to show the benefits of resveratrol with long-term use; indeed, it has not been shown to decrease cellular senescence.

In the next chapter, we build on this understanding of the principles of cellular efficiency and look at how the most important hormones and growth factors that the body produces can influence the maintenance of cellular efficiency.

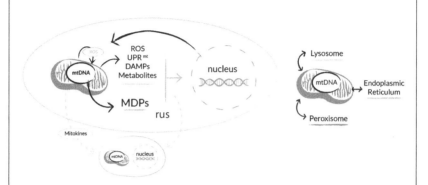

Upregulate the SIRT1 and SIRT3 genes. Stimulate optimal oxygen utilization and production of cellular energy by producing more ATP and generating low ROS. This leads to an optimal balance of AMPK, and mTOR leads to improved autophagy, mitophagy, and protein folding.

4

Harnessing Growth Hormone to Improve Cellular Efficiency

One of the more global effects of senescence and the progression of cellular dysfunction is compounded by a decrease in growth hormone (GH) release and IGF1 production, which are intended to act synergistically to promote overall cell growth and proliferation. In senescence, the reduction of our master hormone and its companion IGF1 trigger increases in cellular cortisol and sarcopenia and a decrease in glucose sensitivity. In addition, with the increased mitochondrial dysfunction, we have decreased cognitive function, a less effective immune system, decreased mitochondrial biogenesis and efficiency, increased reactive oxygen species, and decreased steroid production. These are many of the downstream impacts triggered by the SASP and all can be directly related to decreasing growth hormone release and IGF1 production. Let's take a look at the mechanisms for growth hormone production.

HOW GROWTH HORMONE FUNCTIONS IN THE BODY

Growth hormone is significantly involved in cellular proliferation and efficiency and is also intimately involved in the production of IGF1, which is also involved in cellular growth, repair, and cell survival. As we noted earlier, cellular efficiency is related to a multitude of physiologic factors that have everything to do with an increase in growth hormone release and IGF1. We can upregulate beta oxidation, oxidative phosphorylation, PGC-1alpha, and PPAR-gamma and improve satellite cell activation. We can decrease cellular senescence, cellular apoptosis, and intercellular cortisol production. We can also improve the mechanism of intracellular steroidogenesis (the first step of steroid production occurs in the mitochondria, where cholesterol progresses to pregnenolone).

In particular, having efficient mitochondria means improved steroidogenesis, which is rooted in the physiologic release of growth hormone and IGF1. Over time, as inflammatory disease processes develop and increased cellular senescence occurs, growth hormone and IGF1 become even more important.

Growth hormone is typically released from the anterior pituitary after a signal from the hypothalamus; this signal is a peptide, growth-hormone releasing hormone (GHRH), which stimulates the somatotroph in the anterior pituitary to start the machinery that produces growth hormone. This anterior pituitary somatotroph is also controlled by another peptide called *somatostatin*, which typically inhibits the release of growth hormones and only acts under certain metabolic circumstances or circadian rhythms that assist in decreasing this inhibition.

But as we age, this inhibition becomes more significant. Now, whether that's related to an increase in inflammatory states or increased cellular senescence, that's still up for debate. However, it's my belief that cellular senescence plays a significant role in decreasing the ability of the anterior pituitary somatotroph to go through its typical function of releasing growth hormone. On average, growth hormone is released anywhere from three to six (and even potentially up to nine) times a day in mice, where it is pulsed about every three hours. In humans, it's more like three to six (potentially eight) times a day, where it's also pulsed every three hours;

this pulsing progressively decreases as we age. So, to understand growth hormone further, when it's working optimally, its release happens during the night, with the biggest release of GH occurring in the first sleep phase and then later in stage 4 sleep.

Sleep is important for many physiologic functions, but it's most significant for repair and recovery. As we age, we lose the capacity for stage 4 sleep, and hence we lose the growth hormone response. Stage 4 sleep is also the time when the lymphatic system and glymphatics of the brain are draining. This is an important way that the brain takes care of cleaning up and removing metabolic debris. In this way, GH and GHRH are intimately involved in improving the immune system, structural repair, and the growth and restoration of tissue.

Growth hormone receptors have been identified on most tissues in the body including muscles, adipose tissue, and the liver, the heart, the kidneys, the brain, and the pancreas. The widespread nature of these receptors points to the importance of growth hormone for protein anabolism, promotion of lipolysis, and resistance to insulin-induced glucose metabolism in the liver and peripheral tissues. Growth hormone promotes protein synthesis, typically resulting in a reduction of protein breakdown, hence sarcopenia. In addition, growth hormone improves metabolism efficiency by increasing ATP production. Again, growth hormone is significantly important in promoting an increase in mitochondrial oxidative capacity and in increasing the mitochondrial transcription genes—also features of cellular efficiency. Growth hormone also increases muscle RNA, the encoding of IGF1, and mitochondrial proteins for the nucleus for oxidative phosphorylation.

In turn, growth hormone also upregulates transcription factors and glucose transporters; PGC-1alpha is upregulated, improving its downstream positive effects on the COX3, COX4, TFAM, and GLUT4 messenger RNA during growth hormone release.

Growth hormone typically plateaus in young adult life and then progressively declines, accompanied by a loss of muscle mass and aerobic capacity and an increase in abdominal visceral fat. Typically, after the third decade of life, there's a progressive decline of growth hormone secretion by about 15% every decade of adult life. The secretion of growth hormone at puberty is about 150 micrograms per kilogram per day; it then decreases to about 25 micrograms per kilogram per day by the age of 55. Not only is there a decrease in the pulse amplitude but also potentially a decrease in pulse frequency; a lot of the growth trophic effects of growth hormone are mediated by the downstream production of IGF1, which also declines with aging and parallels decreases in growth hormone.

However, when it comes to helping fight against senescence and promoting cell efficiency, we can look at a centenarian, a hundred-year-old, who can store about the same amount of growth hormone as a 20-year-old in their anterior pituitary. In other words, it's really about the machinery, and its producing and releasing the growth hormone, that becomes problematic. Octogenarians and centenarians have the same capability of releasing growth hormone as young adults, and therefore growth hormone supplementation can have significant and important ramifications for them. It can aid in improving lean body mass, decreasing adipose tissue, and increasing bone mineral density, which is important for decreasing the progression of sarcopenia. We also know that growth hormone has a significant influence on the brain's cognitive functioning, also improving fluid and crystallized intelligence—all of which can be traced back to IGF1 concentrations and production.

We know that, as we age, IGF1 expression in the brain decreases, corresponding to decreasing hippocampal neurogenesis. IGF1 is very important in promoting organized adult hippocampal neurogenesis; it not only promotes adult neogenesis through increased stem cell proliferation, but it also promotes it through organized cell migration. These reduced IGF1 levels are linked to cognitive dysfunction and can be correlated with decreases in motor performance, speed of information processing, and fluid intelligence. IGF1 also works together with brain-derived neurotrophic factor (BDNF) and other neurotrophic factors to promote neurogenesis and remodeling of the brain.

Clearly, growth hormone and IGF1 are important hormones. They promote and maintain cellular survival; improve cellular efficiency, utilization of glucose, and cellular repair; prevent insulin resistance; and promote continued brain neurogenesis and maintenance of hippocampal function for memory consolidation and recall. So why not simply give GH to counteract a loss of production capacity?

Historically we can look at exogenous use of growth hormone and multiple studies that have confirmed using it show a definite improvement in lean body mass, decreased adipose tissue mass, and increased bone density. However, we also know that using GH itself is not a physiologic stimulator of the release of growth hormone. It eliminates negative feedback loops and creates supraphysiologic increases in growth hormone that lead to increases in reactive oxygen species and dysfunction of the mitochondria, and its oxidative effect is diminished. Exogenous GH can also lead to brain receptor involution; specifically, it can cause an increase in anxiety and fear. In fact, the overstimulation that results from using exogenous growth hormone prompts an mTOR dominant state, causing us to lose mitochondrial efficiency, mitochondrial biogenesis, and the capability of the mitochondria to produce its cytochromes, which are important in the electron transport chain. The nucleus still makes its cytochromes, but the loss of mitochondrial cytochromes leads to an inefficiency of the electron transport chain, and thus decreased oxygen efficiency with reduced ATP and increased reactive oxygen species because of the continued mitogenic and proliferous state of the cell. In other words, exogenous GH creates SASP, encouraging more senescent cells to develop, and causing complications from insulin resistance, cardiomyopathies, and other metabolic diseases. Negative feedback loss of GH production leads to supraphysiologic production of IGF1, which causes premature cellular senescence because of the deactivation of the SIRT1 control of p53 guided cell senescence.

So, if GH is not the answer, what is?

Peptides; specifically, a group of peptides that act on the release of growth hormone, hence their names: growth-hormone releasing hormone (GHRH) and growth-hormone releasing peptides (GHRPs).

GHRH/GHRP: HOW YOU CAN RESTORE CELL EFFICIENCY

When we recognize the decline of GH and IGF1, we can strategically improve cell efficiency in a number of ways:

- Optimize beta oxidation of fatty acids

- Optimize the TCA cycle (aka the Krebs cycle)
- Optimize oxidative phosphorylation
- Increase AMPK
- Increase PGC-1alpha
- Increase PPAR-gamma, alpha, decrease TGF-beta
- Increase NRF1,2 antioxidant related elements (AREs)
- Increase TFAM
- Optimize NAD+/NADH pool
    - Optimize SIRT gene activation
    - Optimize FOXO gene activation
- Decrease IL2, IL6, upregulate IL10
- Block nuclear transcription of NF-kB; leads to decreased IL1 Beta, IL6, TNF-alpha, IL18

The following actions can help to reregulate autophagy in beta cells and increase the formation of autophagosomes.

- Restore insulin receptor (IR) signaling, which acts synergistically with GLP-1 signaling and modulates autophagy, oxidative stress, protein synthesis, apoptosis, and mitochondrial biogenesis.

- Alleviate glucotoxicity, lipotoxicity, excess nitric oxide (NO), increased cytoplasmic calcium, oxidative stress, and cytokine-induced endoplasmic reticulum (ER) stress in both primary beta cells.

- Reduce glial inflammation to elevate levels of the inhibitor of NF-ϰB (IϰB-a), ultimately leading to a reduction in neuroinflammation.

- Address mTOR by controlling for geoconversion (i.e., cell committing to senescence); use rapamycin to slow geoconversion and preserve cell proliferative potential.

DELAY SENESCENCE THROUGH GHRP

The GHRPs have two receptors that are important to cell efficiency, besides their direct effect on growth hormone release; this graph emphasizes their pleiotropic effects. Look at how the CD36 receptor is integral in the control of tissue fibrosis, which is late sequelae of senescent cells.

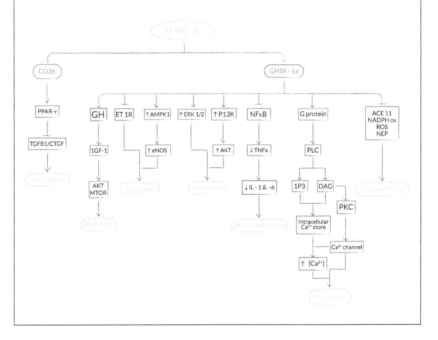

5

Preparing the Immune System in the Age of Viruses, Bacteria, and Other Pathogens

As I write this book, all of us on the front lines of medicine and healthcare are trying to do our best to reassure patients, titrate supplies, and otherwise stay patient, mindful, and clear-eyed as we figure out how best to deal with the ongoing COVID-19 pandemic. Over the past two decades, we have experienced three major worldwide viral pandemics, including the 2001 SARS COV 1 in southern China, the SARS MERS in 2008 from the Middle East, and now COVID-19. Similar to SARS and MERS, COVID-19 is a type of virus that primarily attacks the lungs and upper and lower respiratory pathways, causing an inflammatory response that shuts down varied aspects of the circulatory system. Symptoms include cough, fever, headaches, loss of smell and taste, and inflammation of the toes and feet.

But unlike MERS or SARS, COVID-19 is a novel coronavirus, which means very few people are showing antibodies and not enough time has passed for exposure to expand and create herd immunity. We are also not entirely sure how this particular virus spreads, why certain people test positive for the virus but are asymptomatic, and why still others who are seemingly healthy develop fatal cases of the disease. And although scientists, physicians, and other researchers are working tirelessly to test and ascertain antiviral protocols, there is no vaccine as of yet, and vaccines may or may not provide sufficient protection—at best, current flu vaccines work only 50% of the time. It's important to understand the potential limitations of vaccines to ensure safety.

COVID-19 is showing up in ways that vary across the population, irrespective of age, gender, or ethnicity. Women, men, and children are presenting strokes, heart attacks, acute diabetes, and blood clots. We know that those people with preexisting conditions such as diabetes, heart disease, and COPD are more at risk of experiencing *severe* symptoms of COVID-19 because of compromised cell functioning and tissue degradation in those areas of the body affected by disease. Diabetes, a form of chronic inflammation of the gut, disrupts metabolism, and one end result is a highly acidic intestinal lining, which undermines one of the body's first lines of defense—its microbiome.

People with heart disease, high blood pressure, and atherosclerosis are also more vulnerable to COVID-19. When the heart and its vessels are constricted, the organs require more energy to function. The same is true of lung conditions such as COPD and emphysema. These correlations between underlying conditions and COVID morbidity do share something in common: an immune response that is somehow not working efficiently or sufficiently.

Most simply, we know that any time one part of the body is weak, other parts of the body are affected; a transfer of energy typically occurs so that the energy is directed to the areas where it's most urgently needed. In essence, any virus or bacterial infection is met by the body's immune response, which is made up of a complex interplay of biochemical, metabolic, and cellular reactions designed to fight against a pathogen. The innate system is that set of responses that occur as soon as the body recognizes an antigen's presence in order to prevent infection and isolate or destroy the invader.

It's important to understand the basics of how our immune system works: our bodies are designed to automatically identify foreign invaders. A virus is a foreign invader. One of the first ways our immune systems fight a foreign invader is by sending cell signals to all parts of the brain and body to turn on its defense system. Sometimes this means inflammation—like when you stub your toe and it swells or when you get a cold and your nose gets stuffy. Inflammation is a signal that something is wrong.

It's the *innate immune system*'s job to be the first responder, sending out several signals to attack any kind of pathogen (like a virus or bacteria invader). This frontline defense releases gamma delta T cells that act like an alarm system, signaling a series of reactions. Once this cycle is activated, the *adaptive immune system* assists the innate system to signal the TReg cells to call for macrophages, neutrophils, monocytes, and dendritic cells to engulf and otherwise get rid of the virus or other invader. Further innate responses ensue with assistance from the adaptive immune arm, leading to TReg cells directing further macrophage, neutrophil, and dendritic cells to engulf or neutralize the virus or other pathogens. Other mechanisms are at work as well, enabling the two immune systems to keep interacting efficiently.

Think of the two systems this way: the innate system is on the front line and turns on the invaders with a series of bows and arrows, pushing the virus into retreat mode. Next, the adaptive immune response comes in with its antibodies like an army of Pac-Men, gobbling up the weak, retreating invaders.

Keeping these systems working in tandem to protect and fight against something as nasty as COVID-19 requires that we feed our cells optimally. This is why nutrition, exercise, and high-quality sleep are so important: they give the cells of the immune systems not just adequate sources of energy but preferable nutrition.

These two arms of our immune system modulate each other; they also must work together to fight off any disease or infection. If one arm is compromised, it will trigger the other to work harder. If one is in a hyperactive mode, overworking, then the other arm will begin to dampen its response. The two arms are always in a give-and-take exchange, and if this exchange becomes dysregulated, further problems occur, including respiratory distress, blood clots, stroke, neurological symptoms, and the exacerbation of some autoimmune disorders, such as lupus or rheumatoid arthritis. These are downstream effects that we are seeing in COVID-19 patients.

Right now, the key for us is to give our cells the best possible opportunity to stay efficient so they can remain intelligent and control the processes of better DNA repair, autophagy, mitophagy, and apoptosis, all of which lead to better epigenetic influence, decreasing the senescent secretory phenotype potential of aging and disease. I keep talking about cell efficiency for a reason: so much of our overall health depends upon our body's cells being able to get the optimal nutrition, which leads to NAD+, NADPH, ATP, and Acetyl-CoA production for energy, and utilization of those energy sources to keep active and growing. When this cell cycle gets depleted—from either poor sources of nutrition or an immune response that is diverting nutrition or interrupting energy production or utilization—then cells can become senescent, which is the hallmark of all sorts of downstream diseases.

INNATE AND ADAPTIVE IMMUNE SYSTEMS

The innate immune system is made up of various components, including physical barriers (such as the skin, epithelial and mucous membranes, and mucus itself); anatomical barriers; epithelial and phagocytic cell enzymes (such as lysozyme); phagocytes (neutrophils, monocytes, macrophages); inflammation-related serum proteins (C-reactive protein and lectins, for example); and antimicrobial peptides (including defensins and cathelicidin). The innate system also sends signals to toll-like receptors, which in turn release cytokines and inflammatory mediators (such as macrophages, mast cells, and natural-killer cells).

These mechanisms create a cascade of signaling events to prevent infection, eliminate invader pathogens, and then *turn on* the acquired immune response.

(For a more detailed description and analysis of the innate system, visit www.ncbi.nlm.nih.gov/books/NBK459455/.)

The *adaptive* or *acquired immune system* occurs in conjunction with the innate system, and ideally is responding to signals from the innate system in an ongoing way. Like the innate response, adaptive response is designed to protect against further infection. It relies principally upon two specific cell types—beta cells and T cells—that respond to specific antigens. Again, whereas the innate system's response, including inflammation and the release of cytokines, occurs within minutes or hours of contact with a pathogen, when it's triggered to take over or help the innate system, the adaptive immune response occurs after several days. In general, the adaptive response entails several discrete steps:

1. Recognize the antigen.

2. Release white blood cells (lymphocytes), most significantly to assist in TReg differentiation.

3. Activate and proliferate the responding cells.

4. Transcribe genes.

5. Synthesize proteins.

6. Produce specific end products, such as antibodies, cytokines, and so on.

The immune system is intended to preserve homeostasis by combatting any internal (endogenous) or external (exogenous) stressors, toxins, or other agents that upset its capacity and efficiency for managing allostatic load. A preexisting condition like diabetes, heart disease, or COPD has already increased the allostatic load of the immune system—both its innate system and its adaptive—and therefore all protective or defensive capabilities have become overtaxed and downregulated. This combination, with increased senescence associated secretory phenotype and, as previously described, an overtaxed ACE2 and CD26 receptors, sets the stage for immune dysregulation.

The world is anxiously in search of the appropriate vaccination or antiviral medication for stopping this specific coronavirus. I openly invite you to critically evaluate and consider whether, in conjunction with this pathogen, it is necessary to prepare the cell for ultimate function and efficiency to improve any adjunctive treatment. Cell inefficiency and loss of metabolic flexibility tip the scales for any epigenomic stressor to influence cell phenotype, which further propagates cellular senescence.

The time has come to recognize the metabolic efficiency of the cell as being a significant factor in preparing the cell for ultimate function when it is faced with unpredictable stressors like those present with the COVID-19 pandemic. The growing stressors from increasing aging population to higher rates of metabolic, immunologic, microbial, and oncogenic disease further exacerbate the complications of understanding COVID-19. There's a common theme here: the dysregulation of nucleotide cofactor ratios, which lead to increasing reactive oxygen species (ROS) and reactive nitrogen species (RNS), decreased mitochondrial efficiency, and loss of SIRT and FOXO gene transcription factors, triggers insufficient endogenous antioxidant functioning that is necessary to regulate cellular efficiency. When this occurs, we see enhanced stem cell, immune, and general cellular senescence depending on the environment.

COVID-19 viral penetrance is dependent on ACE2 receptors (1 and 2). As we have already established, disease and aging ACE2 receptors are depleted, and this population is already disadvantaged by the loss of the regulation in the renin angiotensin system, which is dependent on checks and balances between the pro-inflammatory renin/Ang II/AT1R and the anti-inflammatory ACE2/Ang(2 COVID19 through 7)/MAS. Appreciating this dysregulation and further appreciating Ang II imbalance increases TGF-beta1 (TGFB1) and leads to an increased production of interleukin 10 production, which continues a dysregulation of the innate and acquired immune response.

The presence of a low-grade innate progression with increasing TGFB1 and increased signaling of interleukin 10 leads to untimely Interleukin 10 constantly working against the innate system and directly influencing a decrease in gamma delta T lymphocyte activation, which is needed to initiate a first response to the virus. Following this pathway, we add one more benefit of considering metabolic therapy. It has been made evident that sodium butyrate has a direct inhibition on renin production. A metabolic approach that considers the use of exogenous ketones could give a considerable advantage to balancing the RAS where Ang II predominates and is further enhanced by viral uptake of the remaining ACE2 receptors.

Here I am emphasizing how metabolic influences directly and indirectly regulate immune modulation. Metabolic flexibility enables the inherent intelligence of the cell to regulate cellular redox and influence positive epigenetic determinants for a beneficial cell phenotype. Later in this book, I introduce specific immune cell modulating peptides like thymosin alpha 1 (TA1), thymosin beta 4 (TB4), and other peptides that, when combined with the appropriate nutrition and exercise, give individuals the necessary control of preparing their immune system for any potential assault.

Recent COVID-19 research out of China validates the significance of TA1 treatment in reducing mortality in severe COVID-19 patients. TA1 was shown to restore CD4 and CD8 T cells in circulation and reverses CD8 T cell exhaustion, thus reducing PD1 and TIM3 expression. PD1 is a programmed cell death receptor on the CD8 T cell that upregulates and directs attack on normal cells in the body (loss of self-tolerance). TIM3 regulates continued interferon gamma release when activated. Continued activation leads to the innate immune response going into overdrive. In my opinion, this paper represents a tremendous step forward, validating immune modulating peptides like TA1. We now have more information to not only assist patients with immune overdrive, but also to prepare the immune system for improved modulation in preparing for viral exposure to COVID-19 or any other viral pathogen.

Our best defense against any virus, including this coronavirus, is offense. Our bodies are supremely intelligent and learn quickly. If we give our bodies the right information, we can jumpstart our immune systems both to resist COVID-19 and also to repair what's under assault. Viral infections like COVID-19 are here to stay. The best way we can help our patients protect themselves and their loved ones is to continue to support their immune systems. Viruses love to mutate; the best defense is a strategically well-planned offense.

Peptides to Enhance Cellular Functioning, Regulation, and Efficiency to Delay Cellular Senescence

We have thousands of peptides at our disposal, with more compounds being explored and developed all the time. In this section of the book, I've put together five categories of peptides that address the underlying mechanisms of cellular functioning, repair, and efficiency. These peptides offer you a strong baseline for preventing and/or delaying cellular senescence, while also going after inflammation, mitochondrial degradation, and immune system dysfunction.

To recap, cell senescence creates continued loss of cellular homeostasis and creates the environment for negative effects, including these:

- Decreased GH, IGF1
- Decreased mesenchymal stem cell function (loss of quiescence)
- Increased cellular cortisol
- Sarcopenia (catabolic state)
- Decreased SIRT gene activity
- Decreased NAD+/NADH ratio
- Increased NADP+/NADPH ratio
- Loss of FOXO gene activity
- Decreased insulin sensitivity
- Mitochondrial dysfunction
- Decreased cognitive function and memory
- Decreased immune function
- Decreased mitochondrial biogenesis and efficiency with increased ROS
- Decreased steroid production (mitochondria)

These conditions and their root causes can be targeted with specific peptides.

6

Targeting GH and IGF Pathways to Reenter the Cell Cycle

As you know, cell senescence means cell arrest, which means that cells no longer actively participate in the cell cycle. Although there is still considerable controversy on stages of cellular senescence, I am of the opinion that we have more work to do to determine or predict how far a cell can commit to senescence and then still be able to reenter the cell cycle. As I mentioned in the last section, one global way of stimulating cells to reenter the cell cycle is through the GH and IGF pathways and giving the cell the chance to benefit from improved autophagy and mitophagy. As a group, these peptides are referred to as HGH peptides and they work in various ways to improve cell cycle functioning and proliferation to signal endogenous growth hormone. These peptides are also used to improve the landscape for improved DNA repair so the cell can confidently reenter or continue the cell cycle.

The purpose of this group of peptides is to elevate the physiologic release of endogenous GH, improve downstream transcription, and help with the translation of hepatic and, more importantly, extrahepatic cellular IGF1. However, we need to keep in mind that using a GHRH by itself does not necessarily mean there will be an immediate endogenous growth hormone release. The machinery in the anterior pituitary secretagogue is set in motion to produce the pulse of growth hormone, but the hypothalamus still controls the release of GH, with somatostatin having a rate-limiting effect. In this way, though GH will eventually be released, it will not be until somatostatin inhibition is lifted. It's for this reason that it's important to consider using a GHRP in combination to ensure endogenous growth hormone release within a desired 20-minute window (I will go over more specifics on dosage and modes of use shortly).

It's also important to understand how consumed nutritional substrates affect GHRH/GHRP peptides. Pure protein has no effect on endogenous release of GH; on the other hand, carbohydrate and fatty acid consumption *can blunt* the GH release. I recommend following the simple rule of no food for up to 30 minutes after GHRH/GHRP use and no food 1½ hours before use. These are the only peptides now known to be affected by nutrition. A general scheme of use could be utilizing them before bedtime to ensure stage 4 sleep improvement and first thing in the morning before eating breakfast. (I present a general algorithm later at the end of the GHRH/GHRP section.)

When GHRH and GHRP are used effectively, they can upregulate endogenous GH and IGF to improve cell efficiency by

- Upregulating beta oxidation
- Upregulating oxidative phosphorylation
- Upregulating PGC-1alpha
- Upregulating PPAR-gamma
- Improving mitochondrial efficiency
- Upregulating the SIRT gene
- Activating the FOXO gene
- Improving the stem cell stress response and maintaining the quiescent state
- Decreasing cellular senescence

- Improving cellular autophagy
- Improving cellular mitophagy
- Decreasing cellular apoptosis
- Improving intracellular cortisol production
- Improving immune function with a decreased TH17/ TReg ratio
- Improving NAD+/NADH ratio (increased)
- Improving NADP+/NADPH ratio (decreased)

TYPES OF GROWTH-HORMONE RELEASING HORMONE (GHRH) AND GROWTH-HORMONE RELEASING PEPTIDE (GHRP)

Types of GHRH

Sermorelin: First FDA-approved GHRH used to address short stature.

MOD GRF (1-29): The most commonly used, this hormone is known better in the industry as CJC 1295 without DAC (a drug affinity complex that increases the half-life of MOD GRF [1-29]).

CJC 1295

Tesamorelin: This GHRH was FDA approved for visceral adiposity in HIV patients.

Types of GHRP (aka ghrelin mimicking peptide)

Hexarelin: The strongest GHRP in the family; known to give the biggest pulse of all. Hexarelin will create prolactin and cortisol side effects. Desensitization will happen regardless of the dose.

GHRP2: Second strongest in the category; desensitization is unclear if used beyond saturation dose.

GHRP6: Third in the lineup for potency; when shots are broken up, desensitization does not occur. Creates slight prolactin and cortisol issues. GHRP6 is one of the only peptides that is known to actively increase ghrelin in the stomach.

Ipamorelin: This is the mildest of the bunch; it does not create prolactin or cortisol, and at very large doses, it is shown to give a large release of GH without desensitization. (This hormone is the most commonly used.)

MK0677 (Peptide GHRP mimetic): Strong GH and potential supraphysiologic IGF1 response.

HGH vs. Peptides

Peptides offer the same benefits as HGH, without the risks.

	HGH (Somatropin)	HGH Peptides (e.g., Sermorelin)
EFFECT ON HGH LEVELS	Promotes unnatural HGH levels	Promotes natural release of HGH
	Can shut down natural HGH production	Promotes natural HGH production
EFFECT ON PITUITARY GLAND	Can negatively impact pituitary function	Supports pituitary function and health
SAFETY	HGH levels drop when therapy is stopped	HGH production continues for a period even after therapy is stopped
	High risk of overdose	Very low risk of overdose
	Risk of tachyphylaxis	No risk of tachyphylaxis
	Associated with range of side-effects including cancer	Minimal side-effects
ACCESSIBILITY	Controlled substance, hard to access legally	Readily available through legal means
COST	Higher cost ($1,000+ per month)	Lower cost ($200+ per month)
BENEFITS	All the benefits of healthy HGH levels	Same benefits as HGH, without the risks

As mentioned earlier, exogenous GH is not produced simply by stimulating GH, but rather by peptides that stimulate GH-associated pathways. Keep in mind the differences between HGH and GHRH/GHRP.

GHRH PLEIOTROPIC EFFECTS

First discovered in 1965 by Schally and colleagues as glandular extracts, the neurohormone GHRH was later discovered to have a wider role than originally thought. Specifically, in 1981, it was found that pancreatic tumors could produce GHRH. It was then discovered that GHRH receptors exist not only in the pituitary but also in other cells, including

- Stem
- Muscle
- Liver
- Fibroblasts
- Bone
- Fat
- Pancreatic Islet
- Cardiac
- Immune—beta cells/monocytes (not T cells)
- And others

APPLICATIONS:

Today we use GHRH and its analogues to support the cell cycle, including

- Regulation of cell growth
- Proliferation
- Differentiation
- Survival
- Neurochemical regulation of sleep
- Support of deep non-REM sleep (stage 4)
- Support of REM sleep mediated by GH
- Mediation of GABAergic neurons in the anterior hypothalamus/preoptic region

MOD GRF (1-29)

Modified Growth Hormone Releasing Factor (1-29) (MOD GRF [1-29]) is also known as *CJC 1295 without DAC* by compounding pharmacies.

PROPERTIES:

- 29 amino acids: Tyr-D-Ala-Asp-Ala-Ile-Phe-Thr-Gln-Ser-Tyr-Arg-Lys-Val-Leu-Ala-Gln-Leu-Ser-Ala-Arg-Lys-Leu-Leu-Gln-Asp-Ile-Leu-Ser-Arg-NH2
- Molar mass = 3367.89 g/mol
- Analogue of GHRH
- Also known as CJC 1295 *without* DAC
- Half-life is approximately 30 minutes
- Longer acting than sermorelin (GRF 1-29) (5–6 minutes)

APPLICATIONS:

- Bottom line is to improve HGH levels
- Modified in such a way that it makes the pituitary follow a natural pulsatile release manner similar to GRF (1-44), which has a half-life of 5–7 minutes
- Beneficial in promoting muscle growth and fat burning
- Useful in those looking to slow aging
- May improve sleep

DOSAGE:

- A saturation dose of 100 mcg is typically used.
- 1 mg/kg
- Any higher dosage adds minimally to the pulse of GH released.

CJC 1295

CJC 1295 is also known as *MOD GRF (1-29) with DAC* (drug affinity complex)

PROPERTIES:

- Molecular weight (MW) = 3367.9 g/mol
- Sequence: $C_{152}H_{252}N_{44}O_{42}$
- DAC increases half-life.
- Measurable concentration after 6–8 days
- >90% binding to serum albumin
- Elevates GH and IGF1 for several days after a single administration

APPLICATIONS:

- Similar applications to MOD GRF (1-29), with a longer half-life and more potential for increasing IGF1 response above physiologic levels, which may be more advantageous in the short term
- Used for burns or significant soft-tissue injury applications post surgery

DOSAGE:

- Twice a week at 100 mcg, *or*
- 100 mcg daily (This dosage works best for short-term treatment to elevate IGF1 above physiologic levels.)

TESAMORELIN

Tesamorelin is known by the trade name EGRIFTA.

PROPERTIES:

- MW = 5136 g/mol
- GHRH of 44 amino acids

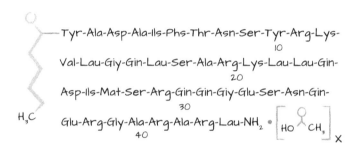

- Developed to treat HIV lipodystrophy
- Pulsed from the anterior pituitary
- IGF1 negative feedback loop is maintained

APPLICATIONS:

- Influences bone mineral density
- Supports muscle metabolism
- Helps lipid metabolism
- Controls inflammation
- Enhances vascular health
- Contributes to improved cardiovascular mortality

When Used for GH Deficiency:

- Higher BMI
- Increased central adiposity
- Higher triglyceride

- Decreased HDL
- Increased hypertension
- Increased carotid intima media thickness (CIMT)
- Elevated CRP

When Used for GH Replacement:

- Increased muscle mass
- Decreased overall fat and visceral adiposity
- Improved dyslipidemia
- Reduced systemic inflammation
- Decreased Charcote-Marie-Tooth Disease (CMT)
- Potential improved cardiovascular health

DOSAGE:

- 2 mg, subcutaneously (Sub Q), daily from studies
- Suggest 1 mg daily. This dose results in a 14% difference in IGF1 stimulation after 6 to 7 days.

POSSIBLE SIDE EFFECTS:

- Increased IGF1
- Injection site erythema
- Injection site pruritis
- Peripheral edema
- Myalgias

GHRP: GHRELIN-LIKE

This peptide is typically used in combination or as a stand-alone because of its function to aid in the immediate release of GH.

PROPERTIES:

Acute ghrelin or agonists (GHRP-2, Ipamorelin, GHRP-6, Hexarelin, MK0677) can

- Be anti-depressive
- Minimize anxiety
- Protect against stress
- Potentially be neurologically protective
- Antagonize somatostatin
- Stimulate release of GHRH
- Increase GH release from somatotrophs in the anterior pituitary
- Ghrelin binds to GHSR (growth hormone secretagogue receptor) to increase GHRH neuron excitability by augmenting their action potential firing rate and decreasing the strength of GABA inhibitory inputs, thereby leading to an enhanced GHRH release.

GHRP Pleiotropic Influences

- Improved GHRH release
- Cytoprotection
- Inflammation/immunity
- Cardiac
- GI
- Brain
- Renal
- Pain

- Arthritis
- Bone mineral density (BMD)
- Neuromuscular

NOTE: It's important to recognize that overstimulation of a GHRP receptor with a high dose of GHRP can desensitize the release of GH by causing an internalization of the GHSR1alpha receptor where the GHSR1beta receptor is activated and causes an internalization of the receptor.

APPLICATIONS:

Ghrelin-like GHRP has a positive effect on

- Cachexia
- COPD
- CVD
- Chronic renal failure
- Chronic respiratory disease

POSSIBLE SIDE EFFECTS:

- Injection site erythema
- Injection site pruritis
- Peripheral edema

GHRELIN/GHRP IN THE BRAIN

Ghrelin in the brain is a stress hormone that acts independent of cortisol. Whether this activity is good or bad depends on how long the GHSR1alpha is activated. (Keep in mind that overstimulation can internalize a receptor with GHSR1beta activation of Hexarelin and MKO677.)

IPAMORELIN

This GHRP is preferred and most commonly used.

PROPERTIES:
- GHRP; third generation
- Increases GH release per somatotrope
- Selective agonist for ghrelin
- Sequence: Aib-His-D-2-Nal-D-Phe-Lys-NH2
- MW = 711.85296 g/mol
- Stable form
- Suppresses somatostatin
- Doesn't raise cortisol, aldosterone, or prolactin levels
- At very large doses, was reported to give a large release of GH without desensitization
- Doesn't promote hunger
- Doesn't have ghrelin's lipogenic properties
- 2-hour half-life
- Increases bone growth
- Improves GI recovery after bowel resection and is a treatment of postoperative ileus

DOSAGE:
- 100 mcg
- 1 mg/kg
- Can be dosed alone or with GHRHs

POSSIBLE SIDE EFFECTS:
- Injection site erythema
- Injection site pruritis
- Peripheral edema

GHRP-6

PROPERTIES:

- Third strongest GHRP
- Hexapeptide
- Sequence: His-D-Trp-Ala-Trp-D-Phe-Lys-NH2
- MW = 873.014 Daltons

APPLICATIONS:

- Used to increase growth hormone (GH)
- Stimulates hunger
- Increases ghrelin
- If cachectic, increases appetite and lean mass
- Can cause a transient increase in cortisol
- Restores GH secretion in obesity
- Improves stage 2 sleep

DOSAGE:

- 100 mcg
- 1 mcg/kg
- Can be dosed with or without GHRH
- Can be used at night to improve sleep
- Can be applied as a microdose at site of injury

POSSIBLE SIDE EFFECTS:

- Injection site erythema
- Injection site pruritis
- Peripheral edema

GHRP-2 (GROWTH-HORMONE RELEASING PEPTIDE 2)

PROPERTIES:

- MW = 817.9
- Sequence: H-D-Ala-D-2-Nal-Ala-Trp-D-Phe-Lys-NH2
- Interacts with ghrelin receptor
- Has a potent stimulatory effect on growth hormone secretion and a slight stimulatory effect on prolactin, ACTH, and cortisol
- Increases growth velocity in children
  - High-dose treatments decrease the levels of both GHRH receptor and GHSR mRNA helping desensitization and downregulation of the receptor
- Improves appetite, weight gain in anorexia
- Normalizes IGF in critical illnesses
- Transient increases in cortisol

DOSAGE:

- 100 mcg daily
- Can be dosed with or without GHRH

POSSIBLE SIDE EFFECTS:

- Increases in cortisol, prolactin, and ACTH
- Increases in appetite, weight gain
- Hypoglycemia

MK0677 (IBUTAMOREN MESYLATE)

PROPERTIES:

- Oral form of GHRP
- MM = 528.7 g/mol
- Sequence: $C_{27}H_{36}N_4O_5S$
- 12.5 mg BID
- Increases GH/IGF1
- Is a non-pulsed ghrelin agonist
- Has a 24-hour half-life
- Can lead to involution of receptors in brain
- It is suggested that this peptide be used no longer than 8 to 12 weeks to avoid potential internalization of the receptor.
- Irreversible neurological damage (prolonged internalization is suggested with no recovery of the receptor)
- Increases cortisol levels (by 2.3 times)

DOSAGE:

The following chart shows a general dosing scheme that is particular to a GHRH/GHRP like MOD GRF (1-29) or Ipamorelin.

Purpose	GHRH/GHRP/BPC	Dosing	Benefits
Introduction 2wks	GHRP	50 mcg Bedtime	Sleep, bone mineral density, well being
Anti-Aging	GHRH/GHRP	100 mcg Bedtime	Sleep, recovery, well being, bone mineral, receptor entraining
Anti-Aging / fat loss	GHRH/GHRP	100 mcg Bedtime 100 mcg Morning	Fatty acid release
Progressive Fat loss	GHRH/GHRP	100 mcg Bedtime 100 mcg Morning 100 cmg 3 hrs later	Fat loss, Enhance fasting
Anabolism	GHRH/GHRP BPC - 157	100 mcg Bedtime 100 mcg Morning 100 cmg PWO 300-600 mcg PWO	Enhanced recovery, ↑GH Receptors
Injury	GHRH/GHRP BPC - 157	100 mcg Bedtime 100 mcg Morning 2 more dosing 3 hr split 100 mcg Split dosing 600 mcg	Recovery, increased healing injury, IGF-1

POSSIBLE SIDE EFFECTS:

- Peripheral edema
- Increase in depression and anxiety can occur with extended use and over-saturation of receptors

EPITHALON (EPITHALON ACETATE TETRA-PEPTIDE)

Epithalon (also known as Epitalon or Epithalone) is the synthetic version of the polypeptide epithalamin, which is naturally produced in humans. This pineal peptide preparation is secreted in the epithalamium-epiphyseal region of the brain. Epithalamin increases a person's resistance to emotional stress and also acts as an antioxidant. It is a bioregulator for the endocrine system, especially for the pineal gland, and has been shown to lengthen telomeres in human cells. It also reduces lipid oxidation and ROS and normalizes T cell function, which helps with cell repair. Additionally, it has restored and normalized melatonin levels in older people who have lost some pineal function due to aging.

PROPERTIES:

- MM = 390.34588 g/mol
- Sequence: Alanine-Glutamate-Asparagine-Glycine
- Anti-aging
- Regulates cell cycle through telomerase activity upregulation

APPLICATIONS:

- Decelerates aging
- Suppresses tumor development
- Induces telomerase activity
- Induces telomere elongation
- Prevents chromosome fusion
- Decreases incidence of spontaneous radiation in carcinogenic tumors
- Normalizes reproductive system in senescent animals
- Improves antioxidant defense
- Normalizes melatonin levels
- Improves cortisol secretion consistent with circadian rhythm
- Improves insulin sensitivity

Effect of Thymalin and Epithalamin on the patients' mortality rate.

INDEX	CONTROL	GROUP				
		Thymalin	Epithalamin	Thymalin + Epithalamin	Thymalin + Epithalamin
			Applied for 2 years			Applied for 6 years
Number of patients		22	24	24	24	20
Baseline age, y. o.		80.2+1.6	80.6+2.5	81.5+2.1	82.1+2.3	79.4+1.8
Age limits, y. o.		70-87	66-93	67-91	67-94	72-91
Mortality rate, % (observed for 6 years)		81.8	41.7*	45.8*	33.3**	20.0**

* - P<0.03, ** - P<0.001, as compared to the control index.

DOSAGE:

- Can be used as a starting therapy to help DNA repair and upregulate antioxidants
- Can be given as a stand-alone therapy twice a year because it's important to cell protection, improving cell resistance, and stopping senescence from occurring
- Can be used intermittently in other scheduled peptide therapies
- 100 mg total; 10 mg IM q day for 10 days, every year, for 2 years
- 50 mg total (Ukraine Academy Medical Sciences); 10 mg IM q every third day, every 6 months, for 3 years

POSSIBLE SIDE EFFECTS:

- Injection site erythema
- Injection site pruritis
- Peripheral edema

7

Cellular Repair: Helping Cells Recover

The peptides in this group are particularly effective in giving cells what they need to restore their functioning after damage. Cellular repair is a complex process that requires efficient cellular metabolism and cell signaling linked to ultimate control by circadian clock mechanisms. Circadian clock control genes, CLOCK, BLAM1 (activation), PER, and CRY (inhibition), influence the master circadian clock in the suprachiasmatic nucleus (SCN) of the anterior hypothalamus. This master clock acts as the pacemaker, directing other independent central nervous system and peripheral tissue oscillators. These genes drive circadian clock oscillations in different biological processes including cellular repair, sleep, locomotor activity, body temperature, hormone levels, and blood pressure. Almost half of the protein coding genes show circadian rhythms in their transcription.

Cell metabolism is regulated by peptides, hormones, enzymes, and transport systems that are directly influenced by circadian rhythm. The master circadian clock influences cells individually, but these cells also have autonomous circadian functions. Modulation of gene expression, secretion of proteins, and activation of metabolites in these cells is defined by circadian rhythms and governed by a network of autoregulatory feedback loops of transcription/translation.

Circadian disruptions are associated with many pathological conditions such as depression, anxiety, pain, fatigue, obesity, diabetes, immune disorders, psychiatric disorders, cancer, and premature aging.

I believe all these pathological entities that are linked to circadian rhythms begin with disruption in cellular repair. Cellular and extracellular maintenance depend significantly on the activity of the fibroblast, and important functions of fibroblasts begin with wound healing. *Fibroblasts* are mesenchymal cells that secrete an extracellular matrix and require motility dependent on actin filaments in the extracellular matrix. F-actin is important for efficient wound healing and motility of fibroblast and most eukaryotic cells. Regulation of the actin cytoskeleton is controlled by actin effector proteins Cofilin-2 and RhoA. The transcription of these proteins is rhythmically expressed. Cytoskeleton changes are necessary for proper cellular adhesion and proper cellular migration is needed for maintenance and repair. If we appreciate the fact that nutrition, sleep, exercise, and life stressors directly influence circadian rhythms, we can then appreciate how just repetitive daily cell maintenance and repair are very dependent on appropriate circadian oscillations. The actin cytoskeleton is essential to cellular division, signal transduction, and efficient cellular maintenance and repair.

Stem cell function is required for daily tissue maintenance and repair. Transcriptional changes under the control of circadian rhythms are essential for the proper function of stem cells. Disrupted circadian clock genes in stem cells demonstrate impaired function and self-renewal.

The peptides that are described in this chapter are the core of many peptides that are utilized for resetting this molecular clock and activating cell signaling pathways necessary for efficient cellular repair.

BPC 157 (BODY PROTECTION COMPOUND 157)

BPC 157 is a 15-chain amino acid peptide that was discovered in and isolated from human gastric juice. It has been shown to accelerate wound healing, including tendon-to-bone and ligament damage. In addition, BPC 157 seems to protect organs and to prevent ulcers of the stomach. This peptide is also shown to decrease pain in damaged areas. It has been found to modulate the serotonergic and dopaminergic systems and offers neuroprotective effects, including neurogenesis, and can work well for patients suffering from traumatic brain injuries (TBI). Working on the brain-gut axis, BPC 157 offers tremendous potential healing for a vast array of cell repair.

PROPERTIES:

- Pentadecapeptide (15-amino acid chain)
- MW = 1419
- Sequence: Gly-Glu-Pro-Pro-Pro-Gly-Lys-Pro-Ala-Asp-Asp-Ala-Gly-Leu-Val
- Focuses on the gut-brain axis
- Human BPC is found in gastric juices.

APPLICATIONS:

- Is used in deep skin burns, corneal injuries
- Is used when there is an injured muscle, tendon, ligament, or bone
- Offers gastric protection
  - Is an antiulcer peptidergic agent
  - Is cytoprotective
  - Improves nitric oxide (NO)
  - Helps improve GI mucosal integrity
  - Decreases the gastric side effects of NSAIDs and alcohol

- Helps heal tissues
  - Reportedly improves cell survival under oxidative stress
  - Increases fibroblast migration and dispersal
  - Induces F-actin formation in fibroblasts
  - Improves angiogenesis
  - Enhances vascular expression of VEGFR2
  - Increases the extent of phosphorylation of paxillin and FAK proteins without affecting the amounts produced
- Neuroprotective
  - Influences serotonergic, dopaminergic, opioid, and GABAergic systems
  - Improves nerve regeneration
  - Decreases neuroinflammation
  - May help with depression
  - Ameliorates alcohol withdrawal symptoms and opposes alcohol intoxication
- Cardioprotective
  - May help regulate blood pressure
  - Rapidly and permanently counteracts the QTc prolongation induced by neuroleptics (such as haloperidol, fluphenazine, clozapine, olanzapine, quetiapine) and prokinetics

DOSAGE:
- Half-life of approximately 4 hours
- For general use 400–600 mcg/day total Sub Q; Oral = 500–1,000 mcg daily
- If injury specific, split dosing into 200–300 mcg BID Sub Q, injected specifically around injury site.

NOTE: BPC 157 counteracts effects corticosteroids have on muscle; results can be spontaneous and improve over 2 to 4 weeks' treatment.

POSSIBLE SIDE EFFECTS:

- Injection site erythema
- Injection site pruritis
- Peripheral edema

GHK-CU (COPPER TRIPEPTIDE GHK-CU)

This naturally occurring copper peptide occurs in human plasma; however, as people age, they lose capacity for production. This tripeptide GHK-Cu helps activate wound healing, regulates immune response, acts as an anti-inflammatory and antioxidant, and stimulates collagen synthesis. Research points to GHK-Cu helping to modulate gene expression with anti-aging benefits.

PROPERTIES:

- MW = 403.9242 g/mol
- Naturally occurring copper complex of a glycyl-L-histidyl-L-lysine peptide
- Has a high affinity for copper
- First isolated from human plasma, but is also found in saliva and urine
- We lose GHK as we age; at age 20, the plasma level of GHK is about 200 ng/ml. By age 60, it declines to 80 ng/ml.
- Decline in GHK coincides with noticeable decrease in the rejuvenative capacity of an organism.

APPLICATIONS:

- Activates wound healing, including gastric
- Attracts immune cells
- Is an antioxidant
- Is an anti-inflammatory
- Stimulates collagen and glycosaminoglycan synthesis in skin fibroblasts
- Modulates the activity of both metalloproteinases and their inhibitors
- Improves stem cells
- Defends against tumors
- Restores replicative vitality to fibroblasts after radiation therapy
- Helps regenerate skin; useful for diabetic skin ulcers
- In cosmetic products, GHK has been reported to
  - Stimulate hair growth
  - Tighten loose skin
  - Improve elasticity
  - Improve skin density and firmness
  - Reduce fine lines and wrinkles
  - Reduce photodamage and hyperpigmentation
  - Increase keratinocyte proliferation
  - Stimulate nail growth

DOSAGE:

- 1 to 2 mg/day for 6-week intervals
- Can be utilized 3 to 4 times a year

POSSIBLE SIDE EFFECTS:

- A possibility of copper toxicity; monitor carefully.
- The lunula of the nail turns blue (corrects over 4 to 6 weeks).

DSIP

As we know, a consistent amount of sleep, including REM and deep sleep (stage 4), is an important predictor of overall immunity and cellular recovery. DSIP (delta sleep-inducing peptide) is a peptide that not only addresses sleep disturbance, but also helps in cellular repair by inducing alpha waves, improving REM, resetting the Circadian Clock genes, and suppressing paradoxical sleep.

PROPERTIES:

- Sequence: N-Trp-Ala-Gly-Gly-Asp-Ala-Ser-Gly-Glu-C
- MW = 849
- Half-life: 7 to 8 minutes
- Naturally occurring
- First isolated in rabbits
- A similar peptide is found in high concentrations in human milk.
- DSIP shows a diurnal pattern.
- Can cross the blood-brain barrier

APPLICATIONS:

- Acts as an anticonvulsant
- Has a neuroprotective effect
- Attenuates emotional and psychological responses to stress
- Attenuates a corticotrophin releasing factor on the pituitary gland
- Has antioxidant benefits that can slow down cell damage
- Decreases excitotoxicity as a result of its influence on the NMDA-subtype of the neuronal glutamate receptors
- Modulates neurotransmitter balance

FACTS ABOUT SLEEP

In order to use DSIP effectively, it's important to keep the sleep cycle in mind.

- There are five sleep stages: 1, 2, 3, 4, and REM (rapid eye movement) sleep.

- Stages 3 and 4 are referred to as *deep sleep*, *slow-wave sleep*, or *delta sleep*.

- The first sleep cycle takes about 90 minutes. After that, they average between 100 to 120 minutes.

STAGE	FREQUENCY (Hz)	AMPLITUDE (MICRO VOLTS)	WAVEFORM TYPE
awake	15-50	<50	alpha rhythm
pre-sleep	8-12	50	theta
1	4-8	50-100	spindle waves
2	4-15	50-150	spindle waves and slow waves
3	2-4	100-150	slow waves and delta waves
4	0.5-2	100-200	
REM	15-30	<50	

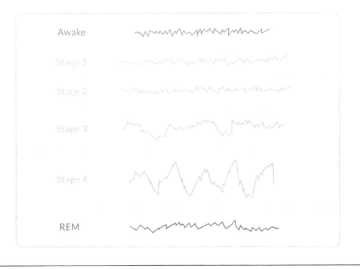

Awake

Stage 1

Stage 2

Stage 3

Stage 4

REM

Slow-Wave Sleep Disruptors

- Sleep deprivation
- Parkinson's disease
- Diabetes and insulin resistance
- Fibromyalgia
- Alcoholism
- Narcolepsy
- Depression
- Anxiety
- OCD
- ADHD

Background

- Delta waves are the predominant wave forms of infants.
- Delta waves have been shown to decrease across the lifespan.
- When a person reaches 75 years of age, stage 4 sleep and delta waves may be absent.

ENDOCRINE FUNCTIONS:

- Reduces plasma ACTH
- Stimulates release of luteinizing hormone (LH)
- Increases GHRH secretion
- Releases thyroid stimulating hormone (TSH) because of being in slow wave sleep phase (SWS)
- Increase in endorphin release centrally to cope with pain
- Restores disrupted sleep

- Assists with alcohol and opioid withdrawal
- Has an antihypertensive effect
- Stimulates antimetastatic activity
- Aids in soothing chronic pain
- Has either a direct or indirect effect on body temperature and alleviates hypothermia
- Enhances neurocognition
- Can increase testosterone levels because it stimulates the release of LH

DOSAGE:
- 100 mcg Sub Q at night, 3 hours before bedtime
- Frequency varies based on clinical response
- Could be given daily, every 3 days, every week
- When the patient stabilizes, doses may be decreased to 50 mcg.

POSSIBLE SIDE EFFECTS:
- May disrupt sleep
- As with all injections, redness and pain at the site of injection may be present.
- Transient headache, nausea, vertigo

NOTE: Naloxone is reported to block the effects of DSIP.

DOSING WHEN USING DSIP/GLYCINE (DELTARAN) (DSIP COMPOUNDED WITH GLYCINE)

- 100 mcg Sub Q
  - Adds to CNS inhibitory effect, similar to GABA, protecting the brain from excitotoxic damage
  - Binds toxic compounds (aldehydes and ketones) produced in large quantities in acute ischemia
  - Upregulates humoral and cellular immunity

TB4 (THYMOSIN BETA 4)

TB4 is an important player in cell repair; I've chosen to introduce it more fully in the next chapter when we discuss immunity, but it's important to keep it in mind when you think of mechanisms for cell repair:

- Is important in wound healing specific to maintaining the actin cytoskeleton with G-actin sequestration; this mechanism is controlled by the circadian clock, which is why DSIP, TB4, and other peptides are administered at night for best cell repair.

- Increases cell motility and migration

- Decreases fibrous growth in ligaments, tendons, and muscles to aid in tissue healing

8

Cellular Enhancement and Immune Modulation

When addressing or anticipating cellular senescence, a frontline offense is to improve the immune modulation between the innate and adaptive cell response. The peptides in this category offer you a one-two punch to enhance overall cellular functioning and improve immune modulation. Your go-to sources for this are the peptides TA1 and TB4. These master peptides work against the inflammatory state of senescent cells. Specifically, TA1 works directly as a senolytic agent that supports apoptosis; TA1 also upregulates glutathione, promotes and improves cellular redox, and initiates IL10 transcription, improving senomodulation. TA1 can take away the camouflage trick that senescent cells use to hide themselves. This then improves the ability of natural killer (NK) cells to eliminate the senescent cell. TB4 is a powerful agent for stopping nuclear factor kappa beta (NF-kB) from transcribing pro-inflammatory cytokines, and so on, thus helping to limit the number of bad messengers the senescence signature sends out.

TA1

Cellular senescence is the primary reason for early dysfunction of the thymus and thymic involution (the shrinking of the thymus with age). As we've seen, cell efficiency depends on optimal mitochondrial function in conjunction with efficient and responsive autophagy. These functions depend on a cell's metabolic flexibility so it can optimally utilize substrates such as fatty acids, glucose, and protein at appropriate times of demand. Loss of cell efficiency due to epigenomic stressors such as poor nutrition, anxiety, viral or bacterial pathogens, or metabolic decline in GH and IGF1, can lead to reactive oxygen species (ROS), DNA damage, and eventual cell senescence. Specifically, cell senescence leads to damaged thymic stromal and thymic epithelial cells (TECs).

In the thymus, the resulting decrease in NAD+/NADH ratios and increase in NADP+/NADPH ratios cause cells to lose the ability to provide sufficient NAD+ for adequate ATP and NADPH production to control free radical production. This process leads to a decline in T cells able to go through differentiation in the bone marrow to become naïve TH0 cells; in the thymus these cells are unable to progress to naïve forms of CD4+ T cells and CD8+ T cells. (Please note: we can also assume that the senescence process influences the bone marrow and because of selection demands, erythrocytes are the primary focus and the differentiation of naïve TH0 cells is also diminished.) The decline in T cells leads to reduced reinforcement to the periphery of the body. The end result is impaired immune response to new pathogens, cancer, auto-immune disorders, increased inflammatory state, and reduced response to vaccinations.

Thymosin alpha 1—or TA1—addresses this confluence of cellular events related to the thymus. TA1 is a major component of Thymosin Fraction 5, a natural thymic peptide that restores the immune function of the thymus and opposes the declines with structural changes of the thymus with aging. Thymosin alpha 1 can increase the Major Histocompatibility Complex 1 (MHC1) on CD+8 T cells and Major Histocompatibility Complex 2 (MHC2) on CD+ 4 helper cells.

TA1 has a pleiotropic effect able to modulate the innate (TH1) and acquired (TH2) immune system to maintain immune homeostasis. TA 1 also acts like a multitasking peptide that can restore overall immune system homeostasis under physiologic and pathophysiologic conditions. It's currently being used to treat Chronic Inflammatory Response Syndrome, Lyme disease, and a multitude of autoimmune diseases; as an adjunct cancer therapy; and as a prophylactic treatment against viral infections, such as COVID-19.

TA1 works on both arms of immune regulation—TH1/TH2. When modulating the TH1 side of the equation, TA1 augments interleukin 2 (IL2), interferon gamma (IFN-g), induction of natural killer cells, and thymopoiesis. TA1 downregulates terminal deoxynucleotidyl transferase (TdT) in the thymus and increases maturation from Th0 T cells to CD+8 T cells and CD+4 T cells. TA1 antagonizes glucocorticoid-induced T cell apoptosis. TA1 can also influence the TH2 arm of the immune system by upregulating indoleamine 2,3 dioxygenase (IDO), which leads to increased FOXO3 and interleukin 10 (IL10) transcription in TReg cells. This effect dampens the TH1 cytokine, chemokine, and protease response on the innate immune system. Glutathione, a powerful endogenous antioxidant, is also upregulated to improve the redox of the cell.

PROPERTIES:

- Synthetic thymic peptide
- 28 amino acids
- Sequence: Ac-Ser-Asp-Ala-Ala-Val-Asp-Thr-Ser-Ser-Glu-Ile-Thr-Thr-Lys-Asp-Leu-Lys-Glu-Lys-Lys-Glu-Val-Val-Glu-Glu-Ala-Glu-Asn-OH
- MW = 3108.28
- Modulates innate immunity
- Pleiotropic

APPLICATIONS:

Cell-Enhancing Applications:

- Promotes T cell differentiation and maturation in vivo and in vitro data
- Decreases T cell apoptosis
- Improves TH1 responses
- Balances TH1/TH2
- Activates indoleamine 2.3-dioxygenase enzyme; dampens immunity
- Enhances dendritic cells
- Enhances antibody responses
- Blocks steroid-induced apoptosis of thymocytes
- Has antitumor effects
- Provides protection against oxidative damage

Disease Applications:

TA1 is often employed in conditions that require immune response modulation.

- For treatment of Hepatitis B and C
- For treatment of HIV/AIDS (can be used in conjunction with oral antiretroviral treatments)
- In cancer treatment as a chemotherapy adjunct
- In non-small-cell lung, hepatocellular, and malignant melanomas
- For treatment of DiGeorge's syndrome
- To improve a depressed response to vaccinations
- For treatment of Lyme disease
- In chemo attractant stimulation
- As an adjunct to flu vaccines, especially in geriatrics

- For treatment of chronic inflammatory/autoimmune conditions (e.g., CFS/fibromyalgia)
- To fight sepsis
- To possibly reduce hematological toxicity of cytotoxic drug therapies, including cyclophosphamide, 5-fluorouracil (5FU), dacarbazine, and ifosfamide

TA1 USED TO TREAT COMMON CHRONIC ILLNESSES

Studies report immune dysfunction is associated with a wide variety of common chronic illnesses; TA1 is used effectively to help treat the following conditions:

- Chronic stress
- Depression
- Metabolic syndrome
- Weight-management issues
- Insulin resistance and type 2 diabetes
- The increased oxidative stress associated with aging
- Chronic fatigue syndrome/fibromyalgia
- Other autoimmune conditions
- Cancer
- Environmental toxins
- Chronic infections, including Lyme, viruses, Candida, and many parasites
- Glutathione depletion that consistently results in a TH1 to TH2 shift
- Dysbiosis
- Food allergies or sensitivities
- Zinc and selenium deficiencies

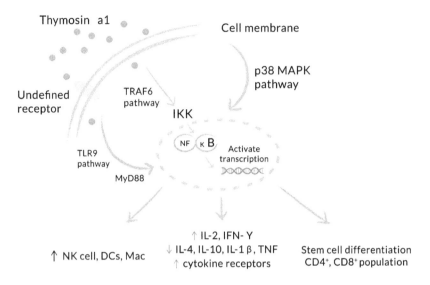

Mode of Action

DOSAGE:

- 2-hour half-life
- 1.5 mg Sub Q every third day
- Treatment from 2 weeks for viral infection to 3 months or longer for HIV, cancer, Hepatitis B/C, or complicated immune suppression or over-activation

ZADAXIN [THYMALFASIN]

This pharmaceutical brand of TA1 has been approved in 30 countries. In the US, it's currently approved by the FDA under the Orphan Drug Designation program. It is currently

- Indicated as a monotherapy or combined with interferon for treatment of Hepatitis B and C
- In Phase III trials for Hep C
- In Phase II trials for Hep B

Dosage:

- 1.6 mg, injected Sub Q, 2 times weekly for 6–12 months
- For patients weighing < 40 kg, adjust the dosage to 40 mcg/kg, 2 times weekly.
- 1.6 mg vials
- Reconstitute with 1 ml = 1.6 mg/ml

Zadaxin has a very low incidence of adverse effects.

TB4

Like TA1, thymosin beta 4 (TB4) is related to *thymosin*, a hormone secreted from the thymus, whose primary function is to stimulate the production of T cells, an important part of the immune system. Thymosin also assists in the development of beta cells to plasma cells to produce antibodies. TB4 is a member of a highly conserved family of actin monomer-sequestering proteins. In addition to its role as a major actin-sequestering molecule, TB4 has a role in tissue repair. TB4 has been found to play an important role in protection, regeneration, and remodeling of injured or damaged tissues. The gene for TB4 has also been found to be one of the first to become unregulated after injuries. TB4 is currently being trialed as a potential therapy for HIV, AIDS, and influenza. TB4 was originally isolated from calf thymus, but since then we have discovered it's more ubiquitous, occurring in most cells.

PROPERTIES:

- Is made up of 43 amino acids: Ac-Ser-Asp-Lys-Pro-Asp-Met-Ala-Glu-Ile-Glu-Lys-Phe-Asp-Lys-Ser-Lys-Leu-Lys-Lys-Thr-Glu-Thr-Gln-Glu-Lys-Asn-Pro-Leu-Pro-Ser-Lys-Glu-Thr-Ile-Glu-Gln-Glu-Lys-Gln-Ala-Gly-Glu-Ser
- This sequence is encoded by the TMSBX4 gene.
- Is produced in the thymus gland
- Is a potent immune modulator
- Exists in higher levels in platelets and white cells
- Upregulates actin
- Is the main intracellular G-actin sequestering peptide
- Forms a ternary complex with actin and profilin

- Mechanisms of Action (MOA): regulates the cell building protein, actin
- Pleiotropic effects: promotes healing and angiogenesis, is a potent anti-inflammatory
- In numerous clinical trials it has been shown to reactivate progenitor cells to repair damaged tissue.
- Promotes rapid wound healing with little to no scarring
- Enhances collagen deposition
- Works at the cellular level supporting tissue stem cells to heal and regenerate the injured tissue
- Works in muscles to protect against sarcopenia as well as post MI
- Promotes angiogenesis and differentiation of endothelial cells
- Is a potent anti-inflammatory for wounds, muscles, and joints
- Reduces acute/chronic pain
- Prevents adhesion and fibrous band formation in injured tissue; i.e., muscles, tendons, and ligaments
- Protects and restores neurons post TBI
- Promotes hair growth

Modes of Action:

- Upregulates cell-building proteins such as actin, a protein that, along with myosin in muscle cells, forms the contractile filaments
- Upregulation of actin allows TB4 to promote healing, cell growth, cell migration, and cell proliferation.
- Due to its low molecular weight (LMW), TB4 is able to "hone" and "travel" to the site of injury.

> **Note:** Some conflicting research has come out about TB4 and cancer:
>
> - Researchers have observed that cancer patients have increased TB4 in affected tissues compared with controls.
>
> - Initially, researchers were concerned that TB4 might be involved in carcinogenesis.
>
> - Presently, the question is this: Is the increase in TB4 a response to cancer? Is it present in the cancer as an immune system response or is it of etiologic concern?

APPLICATIONS:

- Is used as an anti-inflammatory
- Promotes angiogenesis
- Increases collagen deposition
- Is a cytoprotective
- Helps decrease scar tissue formation
- Reduces level of myofibroblasts and supports immunity
- Is an antiviral/antimicrobial
- Improves T cells
- Is used with TA1 as a neuroprotective
- Repairs soft tissues like tendons, ligaments, and muscles from sports/athletic injuries
- Helps treat pressure and venous stasis ulcers
- Treats brain issues if autoimmunity suspected
- Aids in recovery from an ischemic stroke
- Works in muscles to protect against sarcopenia as well as post MI
- Reduces acute/chronic pain
- Improves neuroplasticity
- Repairs and remodels vessels of the heart and other injured tissues

- Can help stem cell differentiation
- Helps patients recover from spinal cord injuries
- When used in conjunction with BPC 157, can help patients diagnosed with TBI/concussion
- Can be employed to help recover from ligamentous, tendon, and muscle injuries
- Has been proven to lessen symptoms of dry eye disorders

NK cell cytotoxicity
↓NF-kB- Nuclear Factor Kappa B
↓Endotoxin lethality
↓Inflammatory
   cytokines: IL-1 βInterleukin
   1 Beta, IL-1a - Interleukin 1
   Alpha, TNFa -Tumor Necrosis
   factor alpha, TXB₂ - T Box
   transcription Factor 2,

MCP-1- Monocyte
Chemotactic Protein 1,
6-keto-PGF1a - Prostaglandin
F2, MIP-2- Macrophage
Inhibitory Factor -2 alpha,
MIP-1- Mac prophage
Inhibitory Factor -1

Adhesion
Migration
Actin-binding
Antimicrobial
Antiapoptotic
MMP-Matrix
Metalloproteinas

TGFβ- Transforming
Growth Factor Beta
Zyxin
N-terminal deoxynu-
cleotidyl transferase

Promotes wound healing

Hair growth
Stem cell maturation
Angiogenesis
Angiopoietin
PAI-1- Plasminogen
VEGF- Vascular
Endothelial Growth Factor

Phagocytosis
Active ILK- Intergrin Linked
Kinase/PCK- protein
kinase C/AKT-Protein
Kinase B

TB4 in action

DOSAGE:

- Between 300 mcg and 1 gram daily, Sub Q
- Depends upon clinical presentation
- Do not dose for more than 3 month.
- Cycle if needed for long-term—3 months on, 1 month off.
- Can be used with TA1 and BPC 157 concurrently
- Individual dosage requirements may vary based on clinical presentation.
- The goal should be to restore TH1/TH2 homeostasis.
- Use to boost NK cells and lower inflammatory cytokines.

POSSIBLE SIDE EFFECTS:

- Reddening, pain, and discomfort at the injection site
- Temporary tiredness or lethargy

MELANOTAN I AND II

These two peptides work on the melanocortin system (MC4R) to increase melanogenesis, which functions as photoprotection/tanning and is important to immune support. In addition, they support the melanocortin system by

- Upregulating TReg cells
- Improving the TH1/TH17 ratio
- Regulating critical inflammation
- Being beneficial in neuro- and cardio-protection
- Supporting autoimmunity
- Targeting sexual dysfunction including improved libido and improved erectile function in men

MODES OF ACTION:

- Binding melanotan I to the MCR1 gene leads to adenylate cyclase (AC) being activated and cyclic adenosine mono-phosphate (cAMP) being stimulated.
- cAMP activates protein kinase A (PKA).
- Results in phosphorylation of cAMP response (CREB)
- Phosphorylated CREB will bind to the cAMP response element (CRE) on the microphthalmia-associated transcription factor (MITF) gene, leading to the synthesis of the MITF protein.
- This results in increased concentrations of the melanogenic enzymes within the melanocyte.
- However, melanotan I doesn't cross the blood-brain barrier, so it does not have any central effects on MC3R and MC4R, NO, sexual desire, and metabolic support.
- Melanotan I (aka afamelanotide, or the brand name Scenesse) is FDA approved for tanning and to prevent phototoxicity in erythropoietic protoporphyria (Dosage: Sub Q implant, 16 mg lasts 2 months).

Melanotan I

Melanotan I is also known as afamelanotide.

PROPERTIES:

- Synthetic alpha-melanocyte stimulating
- Hormone (MSH)
- Sequence: Ac-Ser-Tyr-Ser-Nle-Glu-His-D-Phe-Arg-Trp-Gly-Lys-Pro-Val-NH2
- MW = 1646.845 Daltons
- Non-selective agonist (MCR1)
- Responsible for melanogenesis

APPLICATIONS:

- Sunless tanning effects
- UV protection
- Supports immunity
- Produces photoprotective effects
- Melanocytes favor production of eumelanin (photoprotective black/brown pigment).

DOSAGE:

For Tanning:

For tanning, the dosage depends on patient response and what you're trying to accomplish.

- Use as an "art" for tanning.
- In general, try 200 mcg Sub Q daily for week #1.
- Some patients require 200 mcg daily for weeks, others for only a few days.

- Adjust dosage according to pigment changes.
- After pigment stabilizes, maintain 100 mcg Sub Q twice weekly.

For Immunity:

- 200 mcg daily for 6–8 weeks
- Use until clinical effects become pronounced.
- Very effective with autoimmune conditions
- Sexual stimulation occurs as a side effect.
- Higher doses (i.e., 500–1,000 mcg) can lead to more pronounced sexual stimulation.
- Melanotan sexual stimulation occurs gradually, with long-term benefits.
- PT-141 (bremelanotide) doses twice weekly lead to desensitization more frequently.

Melanotan II

PROPERTIES:

- Stimulates synthetic alpha-melanocyte
- One of the MSH hormones
- Cyclic truncated version of melanotan I
- Sequence:
  Ac-Nle-cyclo[Asp-His-D-Phe-Arg-Trp-Lys]-NH$_2$
- MW = 1024.180 Daltons
- Originally researched for tanning
- Exhibits metabolic support as appetite suppressant, supports glucose and lipid homeostasis
- Is libido-enhancing and improves erection in men

- Promotes skin tanning
- Decreases oxidative stress
- Supports immunity
  - TReg cells
  - Autoimmune support: improves TH1/TH17 balance
  - Stimulates the vagus nerve, which is the cholinergic anti-inflammatory pathway
- Is a neuroprotective—anti-inflammatory

APPLICATIONS:
- Increases melanogenesis
- Increases photoprotection/tanning
- Improves Vitiligo
- Supports the immune system
- Improves TReg cells
- Improves TH1/TH17 ratios
- Supports the autoimmune system
- Used to help with neurodegenerative disorders
- Provides neuroprotection
- Helps fight opiate/ethanol addiction—produces melanocortins involved in decreasing opiate tolerance and decreasing ethanol consumption
- Improves libido and erectile function in men
- Supports appetite and metabolism
- Aids in treating Ischemic diseases, including circulatory shock

DOSAGE:

For Tanning:

- 200 mcg Sub Q daily for 1 week
- Adjust according to pigment changes.
- After pigment stabilizes, 100 mcg Sub Q twice a week, even going to 50 mcg depending on degree of pigmentation

For Metabolic Support:

- 50 mcg daily Sub Q; expect some increased pigmentation

For Immune Support:

- 200 mcg Sub Q daily for 6–8 weeks
- Use until clinical effects are pronounced.

For Sexual Stimulation/ED:

- Use the same dosing schemes as for earlier reasons: sexual stimulation is a desirable side effect!
- In general, higher dosages (500–1,000 mcg) produce more intense effects.
- Sexual stimulation caused by melanotan occurs gradually over time and is more of a long-term effect.

POSSIBLE SIDE EFFECTS:

- Nausea, vomiting, headache, yawning (in up to 10% of patients)
- Do not use if there is a personal or family history of melanoma or non-melanoma skin cancer.
- Melanocortins may increase blood pressure. Use with caution if hypertension is present.
- There is a case report of melanotan II contributing to systemic toxicity and rhabdomyolysis. The dosage in this case was very high (6 mg) and the product was purchased from the internet without a prescription.

- Melanotan II use may result in priapism (an erection lasting longer than 4 hours) in men if recommended dosage is not followed. A case report of a 60-year-old man who was using 10 mg of melanotan II led to a severe case of priapism that required surgery (Winter's shunt) to correct.
- Discontinue use if priapism develops in men.
- It is *not* recommended to use a melanocortin agonist concurrently with a PDE5 inhibitor in men due to risk of priapism.

9

Cellular Efficiency, Metabolic Enhancement, and Weight Loss

covered the role of cell efficiency as it relates to cell metabolism in great detail at the beginning of this book; I encourage you to review that material as needed. In this chapter, I present information related to using specific peptides to target metabolic pathways, specifically mitochondrial functioning, as an additional, effective way to improve cellular efficiency. In particular, the peptides in this section address insulin resistance and other metabolic disorders to improve weight loss, reverse type 2 diabetes, and improve overall energy production and utilization. In this chapter, we will begin by looking at AOD 9604, which is a fragment of growth hormone (GH) and therefore a potent agent for cell efficiency, without any side effects.

AOD 9604

AOD 9604 is a fragment of GH polypeptide (amino acids 176–191) that has been shown to have lipid-reducing effects, similar to but more effective than GH, on account of it not having adverse side effects of unmodified GH. AOD 9604 can regulate fat metabolism by stimulating *lipolysis* (the breakdown or destruction of fat) and inhibits *lipogenesis* (the transformation of nonfat food materials into body fat) both in laboratory testing and in animals and humans. Recent studies also show AOD 9604 possesses other regenerative properties associated with growth hormone. Currently, trials are underway to show the application of AOD 9604 in osteoarthritis, hypercholesterolemia, and bone and cartilage repair.

PROPERTIES:

- Sequence: Tyr-Leu-Arg-Ile-Val-Gln-Cys-Arg-Ser-Val-Glu-Gly-Ser-Cys-Gly-Phe
- MW = 1815.1

MODES OF ACTION:

- Stimulates bone differentiation and mineralization in adipose-derived mesenchymal stem cells (MSC)
- Promotes myoblast differentiation
- Promotes chondrocyte production of collagen and proteoglycan
- Stimulates stem cell differentiation toward bone, muscle, and cartilage repair

APPLICATIONS:

- Unable to induce dimerization and thereby activation of the receptor (no competition with HGH)
- Tyrosin (TYR) in AOD maintains stability of the amino acid sequence; this fragment holds the fat-reducing and tissue repair sequence and mimics the effect of HGH on

lipid metabolism, without having growth-promoting or pro-diabetic effects.

- Inhibits lipoprotein lipase activity in adipose tissue, stimulating lipolysis in adipocytes

DOSAGE:

AOD:

- 250 mcg BID for fat loss
- Sub Q, oral, and topical
- Length of treatment varies; from 6 weeks, to 3 months, up to 6 months

Possible Side Effects:

- No allergenic reactions
- GRAS in foods under the conditions of intended use of AOD 9604

NOTE: Frag 176–191 is a knockoff of AOD; it's supposed to represent an AOD fragment, but it doesn't have a di-sulfide bond to promote stability, which makes it less effective.

AOD/HA

AOD/HA is a combination of AOD 9604 with an added hyaluronic acid (HA), which helps induce cartilage regrowth with intra-articular injection. Further, AOD with HA has been shown to increase IGF1 and improve intra-articular peptide bioavailability. GH can act directly on the growth plate by stimulating local production of IGF1 and by increasing cartilage metabolism and chondrocyte proliferation.

DOSAGE:

- 0.50–1.0 ml used intra-articularly
- Depends on injury and site
- Can be a single injection or split into more than one

For Arthritis:

- 0.50 ml, once weekly, for 5 doses, then .50 ml, once monthly, for 3 doses (total of 8 doses)

GLP-1 LIRAGLUTIDE

Glucagon-like peptide-1 (GLP-1) is a 30-amino-acid-long peptide hormone derived from the tissue-specific posttranslational processing of the proglucagon gene. It is produced and secreted by intestinal entero-endocrine L cells and certain neurons within the nucleus of the solitary tract in the brainstem when food is consumed. The initial product, GLP-1 (1–37), was susceptible to amidation and proteolytic cleavage, which gave rise to the two truncated and equipotent biologically active forms, GLP-1 (7–36) amide and GLP-1 (7–37). Active GLP-1 comprises two alpha helices from amino acid position 13–20 and 24–35, separated by a linker region.

Similar to glucose-dependent insulinotropic peptide (GIP), GLP-1 is the only known incretin that describes its ability to decrease blood sugar levels in a glucose-dependent manner by enhancing the secretion of insulin. In addition to the insulinotropic effects, GLP-1 has been associated with numerous regulatory and protective effects. Unlike GIP, the action of GLP-1 is preserved in patients with type 2 diabetes, and substantial pharmaceutical research has therefore been directed toward developing GLP-1-based treatment.

However, endogenous GLP-1 is rapidly degraded, primarily by dipeptidyl peptidase-4 (DPP-4), but also by neutral endopeptidase 24.11 (NEP 24.11) and renal clearance, resulting in a half-life of approximately 2 minutes. Consequently, only 10–15% of GLP-1 reaches circulation intact, leading to fasting plasma levels of only 0–15 pmol/L. To overcome this, GLP-1 receptor agonists and DPP-4 inhibitors have been developed to resist and reduce this activity, respectively. As opposed to common treatment agents such as insulin and sulphonylurea, GLP-1-based treatment has been associated with weight loss and lower hypoglycemia risks, two very important aspects of a life with type 2 diabetes.

PROPERTIES:

- MW = 3295.65
- Purity ≥ 95%
- Sequence: His-Ala-Glu-Gly-Thr-Phe-Thr-Ser-Asp-Val-Ser-Ser-Tyr-Leu-Glu-Gly-Gln-Ala-Ala-Lys-Glu-Phe-Ile-Ala-Trp-Leu-Val-Lys-Gly-Arg-NH2

DOSAGE:

- Begin 0.6 mg once daily; increase the dose by 0.6 mg weekly to reach a target dosage of 3 mg once daily.
- Doses up to 1.8 mg are used to treat type 2 diabetes. Higher doses (>3 mg) have been shown to decrease body weight by up to 15%.

POSSIBLE SIDE EFFECTS:

- Nausea, diarrhea, vomiting, decreased appetite, indigestion, and constipation

MOTS-c

MOTS-c is a peptide of 16 amino acids produced by a mitochondrial gene, and it has been shown to play a key role in signaling and energy production. Specifically, MOTS-c has been shown to regulate metabolic functions throughout the body, including turning glucose into usable energy. In mice, MOTS-c helped boost glucose metabolism even when the mice were fed a high fat diet. These preliminary studies show evidence for improved control over blood sugar levels for those with type 2 diabetes and obesity.

MOTS-c is also used to increase performance in athletes.

Skeletal muscle has been shown to be the main target organ. Due to its pharmacological effects of regulating metabolic homeostasis, especially the stimulation of glucose uptake and clearance as well as the activation of fatty acid metabolism, MOTS-c can be regarded as a potential exercise mimetic agent and insulin sensitizer.

PROPERTIES:

- MOTS-c is a 16-amino-acid peptide encoded in the mitochondrial genome.
- Sequence: Met-Arg-Trp-Gln-Glu-Met-Gly-Tyr-Leu-Phe-Tyr-Pro-Arg-Lys-Leu-Arg
- MW = 2288.6 g/mol

APPLICATIONS:

- Assists in mitochondrial biogenesis
- Activates AMPK
- Restores homeostasis by initiating catabolic processes for ATP production
- Decreases insulin resistance
- Increases GLUT4 uptake in muscle
- Improves athletic performance
- Improves weight loss

DOSAGE:

- Initially, 5 mg, Sub Q, three times a week. Keep a M, W, F schedule for 4–6 weeks. Follow this with a weekly dose of 5 mg for 4 weeks.

- This treatment can be cycled with other mitochondrial peptides, typically in cycles of 2–3 months.

10

Improving Cell Signaling for Enhanced Cognitive Functioning and Neuroplasticity

For optimal functioning, the brain relies on trillions of neuronal connections made up of neural synapses where electrical and chemical signaling occur. A growing group of peptides are being used to enhance cognitive functioning and offset loss of cognition associated with dementia and other neurodegenerative disorders. The peptides identified here are at the front lines of creating measurable improvements in the brain and nervous system. Also included here are peptides that address pain and enhance neuroprotection.

CEREBROLYSIN

Cerebrolysin, a synthetic nootropic, helps to regulate energy metabolism, can act as a neuromodulator, and can stimulate neurotrophic activity. Essentially, Cerebrolysin protects neurons from free radicals and oxidative stress, acidosis (lactic and keto), and the neurotoxic effects of glutamate. It has also been shown to improve metabolic activity of neurons and enhance cognitive function, memory, creativity, and motivation.

PROPERTIES:

- As a synthetic peptide, Cerebrolysin is a combination of active peptide fragments, including nerve growth factor, BDNF, ciliary nerve growth factor, P-21, enkephalins, and orexin.

- Known to be neuroprotective/neuroregenerative; has neurotrophic repair properties similar to nerve growth factors (NGF) and brain-derived nerve growth factors (BDNF)

- Has a low molecular weight (LMW) so it can cross the blood-brain barrier and the blood-cerebrospinal fluid barrier (BBB/B-CSF)

- Improves synaptic functioning and reduces amyloid deposition

- Decreases amyloid deposition. These effects are accompanied by a reduction in perivascular microgliosis and astrogliosis and increased expression of markers of vascular fitness such as CD31 and ZO-1.

- Reduces the levels of phosphorylated amyloid precursor protein (APP) and the accumulation of APP in the neuritic processes

- Increases the expression of the BBB-GLUT1 via mRNA stabilization

- Markedly increases the uptake of 3H-2-deoxy-d-glucose (essential brain nutrient) and the levels of the GLUT1 protein. Glutamate receptor subunit 1 (GluR1) is one of the four possible subunits of the AMPA-type glutamate receptor. The integrity of this receptor is crucial for learning processes.

- Increases GluR1 density in most measured regions of the hippocampal formation in a highly significant way. These results correlate with the behavioral outcome, revealing an improvement in learning and memory.

APPLICATIONS:

- Helps treat concussion/CTE, TBI
- Used to treat Alzheimer's dementia, mild cognitive impairment (MCI)
  - Decreases beta-amyloid deposition
  - Decreases tau protein phosphorylation
  - Increases synaptic density
  - Restores neuronal cytoarchitecture
  - Results in improved cognitive and behavioral performance
- Employed in treating CVA, TIA
- Used to help with mood dysregulation

DOSAGE:

- 215 mg (1cc) IM/Sub Q daily, 7 days a week, for a 4-week cycle
- 215 mg/ml—5cc IV drip 2 times a week for 2 weeks

SELANK

Selank, previously called TP-7, is a synthetic peptide derived by combining the sequence of tuftsin with another sequence to improve its stability. Tuftsin makes up one part of the immunoglobin G (IgG) antibody and is naturally occurring. On top of its antianxiety, antidepressant, and antiasthenic properties, it can also enhance memory and cognitive function. This peptide can potentially replace stimulants, tranquilizers, and antidepressants.

PROPERTIES:

- Sequence: Thr-Lys-Pro-Arg-Pro-Gly-Pro
- MW = 751.89 g/mol
- Synthetic analogue of human tuftsin

MODES OF ACTION:

- Mimics some of the effects of tuftsin
- Modulates Interleukin-6
- Balances T cell cytokines
- Elevates BDNF in hippocampus
- Has an influence on monoamine neurotransmitters
- Reduces the breakdown of enkephalins

APPLICATIONS:

- Can be used as an antidepressant and an anxiolytic
- Can be used to help treat generalized anxiety disorder (GAD)
- Increases the inhibitory action of GABA
- Improves sleep balance (sleep vs. wakefulness)
- Shows some antiviral activity
- Can help regulate inflammation

- Regulates BCL6 protein, an important transcriptional regulator of the immune system

For Metabolic Syndrome/Weight Gain:

- Prevents weight gain
- Works as an anticoagulant, a fibrinolytic, and an antiplatelet
- Decreases blood glucose levels
- Semax (discussed next) and Selank have anticoagulant and hypoglycemic effects with the same sequence: Pro-Gly-Pro.

For Cancer:

- Reduced tumor of an experimental model in breast cancer

Other Applications:

- Can be used as an anti-ulcer treatment
- Accelerates gastric ulcer healing (ulcers caused by ethanol, stress induced)
- Improves blood supply and lymphatic circulation to the gastric mucosa

DOSAGE:

- Dose 750–1,000 mcg intranasally
- Dose 100 mcg–300 mcg Sub Q daily
- More than these maximum dosages can lead to desensitization
- Dose depends on patient response
- Can alternate with Semax

SEMAX

Semax (N-Acetyl) counteracts the inhibition of learning and memory induced by heavy metals; it counteracts neurotoxic effects; and it inhibits neurodegeneration that is caused by dopamine oxidation. Studies have shown that Semax promotes the survival of neurons during hypoxia and glutamate neurotoxicity, increases the amount and mobility of immune cells, and enhances the expression of chemokine and immunoglobulin genes. In a recent study on brain focal ischemia, Semax influenced the expression of genes that promote the formation and functioning of the vascular system. Overall, Semax is a neuroprotective and contributes to mitochondrial stability under stress induced by the deregulation of calcium ion flow.

PROPERTIES:

- Sequence: Met-Glu-His-Phe-Pro-Gly-Pro
- Is a fragment of the adrenocorticotropic hormone (ACTH, ACTH4-10)
- MW = 813.920 g/mol

MODES OF ACTION:

- Elevates expression of BDNF and the TrkB receptor
- Activates dopaminergic and serotonergic stems
- Can work as an antidepressant and an anxiolytic
- Attenuates chronic stress effects
- Is a potential melanocortin antagonist (MC3R, MC4R)

APPLICATIONS:

- May be used to treat strokes and transient ischemic attacks
- Also used to help with memory and cognitive disorders

- Boosts immune system
- Helps treat peptic ulcers
- Supports the optic nerve

DOSAGE:
- 750–1000 mcg intranasally
- 100–300 mcg Sub Q daily
- More than these maximum dosages can lead to desensitization
- Dose depends on patient response
- Can be alternated with Selank

DIHEXA

This peptide works with hepatocyte growth factor (HGF), which is found in elevated levels in the cortex in early states of neurodegenerative diseases such as Parkinson's. Dihexa has a dual protective effect by increasing HGF activity and lowering dimerization, suggesting that the allosteric modulation of HGF is producing an active monomer complex, theoretically doubling the capacity of available factors to promote signaling cascades and exert changes in cell development. In short, Dihexa has been shown to prevent the development of Parkinson-like symptoms and restore motor function.

PROPERTIES:

- Sequence: Hexanoyl-Tyr-Ile-Ahx-NH$_2$
- MW: 504.66
- Enhances post-synaptic current
- Increases cerebral blood flow
- Encourages synaptogenesis
- Augments spinogenesis
- Stimulates dendritic arborization
- Contributes to calcium signaling

APPLICATIONS:

- Works in the renin-angiotensin system (RAS), which is important for learning
- Aids in memory, neural differentiation, and neural regeneration
- Acts like angiotensin IV (AT4) hexapeptide
  - Known as an analogue of HGF/c-Met
  - Facilitates memory acquisition and consolidation

- Works against the negative influence created by the down-stream effects of octapeptide angiotensin II interactions with the AT1 receptor, which is bad for memory acquisition and consolidation

DOSAGE:
- Topical, 20 mg/ml
- Apply twice a day.
- Use no longer than 6 weeks for neurocognitive enhancement.
- Alternate with Semax/Selank and Cerebrolysin, every 6 weeks.

FGL(L)

FGL(L) peptide is a variant of the natural neural cell adhesion molecule (NCAM) and is known to have neurotrophic and memory enhancing properties. Created directly as a fibroblast growth factor receptor agonist, FGL activates FGFR1 signaling pathways, increasing neurite outgrowth and survival, which in turn leads to memory enhancement. FGL has also been found to improve healing of neuronal tissues by decreasing oxidative stress-induced cell death, inhibiting neuronal degeneration and death.

PROPERTIES:

- MW = 3432.62 g/mol
- Sequence: Glu-Val-Tyr-Val-Val-Ala-Glu-Asn-Gln-Gln-Gly-Lys-Ser-Lys-Ala

APPLICATIONS:

- Improves synaptic plasticity, learning, and memory
- Enhances growth factor–mediated signaling
- Is essential for both early synaptogenesis and synaptic maturation
- Influences the strength of excitatory synapses in an activity-dependent manner
- In cell remodeling and growth, are mediated by fibroblast growth factor receptors (FGFRs)

DOSAGE:

- 1–2 mg daily, for 4–6 times a week, 5 days on, 2 days off
- Younger people in their 30s do better at a lower dose, just 1 mg (they should use for 6 weeks and see how it goes).

BIBLIOGRAPHY

INTRODUCTION

Dawkins, R. (2016, June 8). Epilogue to the mouse's tale—On "epigenetics." Richard Dawkins Foundation for Reason and Science. Retrieved from https://www.richarddawkins.net/2016/06/epilogue-to-the-mouses-tale-on-epigenetics/

Laland, K., Uller, T., Feldman, M. W., Sterelny, K., Müller, G. B., Moczek, A., . . . Odling-Smee, J. (2015, August 22). The extended evolutionary synthesis: its structure, assumptions and predictions. *Proceedings of the Royal Society B: Biological Sciences 282*(1813). doi:10.1098/rspb.2015.1019

One of Darwin's evolution theories finally proved. (2020, March 17). *ScienceDaily*. Retrieved from https://www.sciencedaily.com/releases/2020/03/200317215626.htm

CHAPTERS 1-3

Cell Efficiency and Delaying Senescence

Baar, M. P., Brandt, R. M. C., Putavet, D. A., Klein, J. D. D., Derks, K. W. J., Bourgeois, B. R. M., … de Keizer, P. L. J. (2017, March). Targeted apoptosis of senescent cells restores tissue homeostasis in response to chemotoxicity and aging. *Cell, 169*(1), 132–147.e16. doi:10.1016/j.cell.2017.02.031

Childs, B. G., Gluscevic, M., Baker, D. J., Laberge, R. M., Marquess, D., Dananberg, J., & van Deursen, J. M. (2017). Senescent cells: an emerging target for diseases of ageing. *Nature reviews. Drug discovery, 16*(10), 718–735. doi.org/10.1038/nrd.2017.116

Rowlands, J., Heng, J., Newsholme, P., & Carlessi, R. (2018, November 23). Pleiotropic effects of GLP-1 and analogs on cell signaling, metabolism, and function. *Frontiers in Endocrinology*. doi:10.3389/fendo.2018.00672

Salama, R., Sadaie, M., Hoare, M., & Narita, M. (2014). Cellular senescence and its effector programs. *Genes & Development, 28*(2), 99–114. doi:10.1101/gad.235184.113

Son, J. M., & Lee, C. (2019, January 31) Mitochondria: multifaceted regulators of aging. *BMB Reports 52*(1): 13–23. doi:10.5483/BMBRep.2019.52.1.300

Vizioli, M. G., Liu, T., Miller, K.N., Robertson, N. A., Gilroy, K., . . . Adams, P. D. (2020, January 30). Mitochondria-to-nucleus retrograde signaling drives formation of cytoplasmic chromatin and inflammation in senescence. *Genes & Development 34*(5–6):428–445. doi:10.1101/gad.331272.119

Yang, N., & Sen, P. (2018). The senescent cell epigenome. *Aging, 10*(11), 3590–3609. doi:10.18632/aging.101617

Cell Redox

Elhassan, Y.S., Kluckova, K., Fletcher, R.S., Schmidt, M. S., Garten, A., . . . Lavery, G. G. (2019, August 13). Nicotinamide riboside augments the aged human skeletal muscle NAD+ metabolome and induces transcriptomic and anti-inflammatory signatures. *Cell Reports 28*(7): 1717–1728.e6. doi:10.1016/j.celrep.2019.07.043

Eriksson, S. E., Ceder, S., Bykov, V. J. N., & Wiman, K. G. (2019, April). p53 as a hub in cellular redox regulation and therapeutic target in cancer. *Journal of Molecular Cell Biology 11*(4), 330–341. doi:10.1093/jmcb/mjz005

He, L., He, T., Farrar, S., Ji, L., Liu, T., Ma, X. (2017, November 17) Antioxidants maintain cellular redox homeostasis by elimination of reactive oxygen species. *International Journal of Experimental Cellular Physiology, Biochemistry, and Pharmacology 44*(2): 532–553. doi:10.1159/000485089

Maillet, A., & Pervaiz, S. (2012, April 16). Redox regulation of p53, redox effectors regulated by p53: A subtle balance. *Antioxidants & Redox Signaling, 16*(11), 1285–1294. doi:10.1089/ars.2011.4434

SIRT Proteins

Braidy, N., Jayasena, T., Poljak, A., Sachdev, P. S. (2012, May). Sirtuins in cognitive ageing and Alzheimer's disease. *Current Opinion in Psychiatry 25*(3):226–230. doi:10.1097/YCO.0b013e32835112c

Hsu, Y.-C., Wu, Y.-T., Tsai, C.-L., & Wei, Y.-H. (2018, March). Current understanding and future perspectives of the roles of sirtuins in the reprogramming and differentiation of pluripotent stem cells. *Experimental Biology and Medicine, 243*(6), 563–575. doi:10.1177/1535370218759636

Kilic, U., Gok, O., Erenberk, U., Dundaroz, M. R., Torun, E., . . . Dundar, T. (2015, March 18). A remarkable age-related increase in SIRT1 protein expression against oxidative stress in elderly: SIRT1 gene variants and longevity in human. *PLoS One 10*(3): e0117954. doi:10.1371/journal.pone.0117954

Luo, X.-Y., Qu, S.-L., Tang, Z.-H., Zhang, Y., Liu, M.-H., . . . Jiang, Z.-S. (2014, November 1). SIRT1 in cardiovascular aging. *Clinica Chimica Acta: International Journal of Clinical Chemistry 437*:106–14. doi: 10.1016/j.cca.2014 .07.019

Yamaç, A. H., Kılıç, Ü. (2018, April). Effect of statins on sirtuin 1 and endothelial nitric oxide synthase expression in young patients with a history of premature myocardial infarction. *Turk Kardiyoloji Dernegi Arsivi 46*(3): 205–215. doi:10.5543/tkda.2018.32724

AMPK

Ahmad, B., Serpell, C. J., Fong, I. L., & Wong, E.H. (2020, May 8). Molecular Mechanisms of Adipogenesis: The Anti-adipogenic Role of AMP-Activated Protein Kinase. *Frontiers in Molecular Biosciences 7*: 76. doi:10.3389/fmolb .2020.00076

Kim, J., Yang, G., Kim, Y., Kim, J., & Ha, J. (2016, April 1). AMPK activators: mechanisms of action and physiological activities. *Experimental & Molecular Medicine 48*(4): e224. doi:10.1038/emm.2016.16

FOXO Proteins

Webb, A. E., Brunet, A. (2014, April) FOXO transcription factors: key regulators of cellular quality control. *Trends in Biochemical Sciences 39*(4): 159–169. doi:10.1016/j.tibs.2014.02.003

CHAPTER 4

GH Fundamentals

Frohman, L.A., Szabo, M. (1981). Ectopic production of growth hormone-releasing factor by carcinoid and pancreatic islet tumors associated with acromegaly. *Progress in Clinical and Biological Research 74*:259–71.

Kiaris, H., Schally, A. V., Kalofoutis, A. (2005). Extrapituitary effects of the growth hormone-releasing hormone. *Vitamins and Hormones 70*:1–24. doi:10.1016/S0083-6729(05)70001-7

Rochira, V., & Guaraldi, G. (2017, February) Growth hormone deficiency and human immunodeficiency virus. *Best Practices & Research: Clinical Endocrinology & Metabolism 31*(1):91–111. doi:#xad;label>10.1016/j.beem.2017.02.006

Schally, A.V., Steelman, S. L., & Bowers, C.Y. (1965, May 1). Effect of hypothalamic extract on release of growth hormone in vitro. *Proceedings of the Society*

for Experimental Biology and Medicine 119: 208–12. doi:10.3181/00379727
-119-30138

Thounaojam, M, Powell, F. L., Patel, S., Gutsaeva, D. R., Tawfik, A., . . . Bartoli,
M. (2017, December 12). Protective effects of agonists of growth hormone-
releasing hormone (GHRH) in early experimental diabetic retinopathy.
Proceedings of the National Academy of Sciences of the United States of America
114(50):13248–13253. doi:10.1073/pnas.1718592114

GHRH Influence

Cai, R., Schally, A. V., Cui, T., Szalontay, L., Halmos, G., . . . Zarindi, M. (2014
February). Synthesis of new potent agonistic analogs of growth hormone-
releasing hormone (GHRH) and evaluation of their endocrine and cardiac
activities. *Peptides 52*:104–112. doi:10.1016/j.peptides.2013.12.010

Fridyland, L. E., Tamarina, N. A., Schally, A. V., & Philipson, L. H. (2016).
Growth hormone-releasing hormone in diabetes. *Frontiers in Endocrinology*
(Laussane) 7:129. doi:10.3389/fendo.2016.00129

Kanashiro-Takeuchi, R. M., Tziomalos, K., Takeuchi, L. M., Treuer, A. V.,
Lamirault, G., . . . Hare, J. M. (2010, February 9). Cardioprotective effects
of growth hormone-releasing hormone agonist after myocardial infarction.
Proceedings of the National Academy of Sciences of the United States of America
107(6): 2604–09. doi:10.1073/pnas.0914138107

Obal Jr., F., & Krueger, J. M. (2004, October). GHRH and sleep. *Sleep Medicine*
Reviews 8(5):367–77. doi:10.1016/j.smrv.2004.03.005

Van Cauter, E., Plat, Scharf, M. B., Leproult, R., Cespedes, S., L'Hermite-
Baleriaux, M. & Copinschi, G. (1997, August). Simultaneous stimulation of
slow-wave sleep and growth hormone secretion by gamma-hydroxybutyrate
in normal young men. *Journal of Clinical Investigation 100*(3): 745–753.
doi:10.1172/JCI119587

Xiangyang X., Tao, Q., Ma, Q., Chen, H., Wang, J., & Yu, H. (2016, September
27) Growth hormone-releasing hormone and its analogues: significance
for MSCs-Mediated Angiogenesis. *Stem Cells International 2016*. doi:
10.1155/2016/8737589

GHRH Tesamorelin

Clemmons, D. R., Miller, S., & Mamputu, J.-C. (2017, June 15). Safety and metabolic effects of tesamorelin, a growth hormone-releasing factor analogue, in
patients with type 2 diabetes: A randomized, placebo-controlled trial. *PLOS*
ONE 12(6), e0179538. doi:10.1371/journal.pone.0179538

Falutz, J. C., Mamputu, J-C., Potvin, D., Moyle, G., Soulban, G., . . . Grinspoon, S. (2010, September). Effects of tesamorelin (TH9507), a growth hormone-releasing factor analog, in human immunodeficiency virus-infected patients with excess abdominal fat: a pooled analysis of two multicenter, double-blind placebo-controlled phase 3 trials with safety extension data. *Journal of Clinical Endocrinology and Metabolism* 95(9): 4291–4304. doi:10.1210/jc.2010-0490

Sattler, F. R., He, J., Letendre, S., Wilson, C., Sanders, C., . . . McCutchen, J. A. (2015 March). Abdominal obesity contributes to neurocognitive impairment in HIV-infected patients with increased inflammation and immune activation. *Journal of Acquired Immune Deficiency Syndromes* 68(3), 281–288. doi:10.1097/QAI.0000000000000458

Stanley, L., Chen, C. Y., Branch, K. L., Makimura, H., & Grinspoon, S. K. (2011, January). Effects of a growth hormone-releasing hormone analog on endogenous GH pulsatility and insulin sensitivity in healthy men. *Journal of Clinical Endocrinology & Metabolism* 96(1): 150–58. doi:10.1210/jc.2010-1587

Stanley, T. L., Feldpausch, M. N., Oh, J., Branch, K. L., Lee, H., Torriani, M., & Grinspoon, S. (2014, July 23-30) Effect of tesamorelin on visceral fat and liver fat in HIV-infected patients with abdominal fat accumulation: a randomized clinical trial, *JAMA 312*(4): 380–389. doi:10.1001/jama.2014.8334

MOD GRF and Sleep

Obal Jr., F., & Krueger, J. M. (2004, October). GHRH and sleep. *Sleep Medicine Reviews* 8(5):367–77. doi:10.1016/j.smrv.2004.03.005

CJC 1295

Teichman, S. L., Neale, A., Lawrence, B., Gagnon, C., Castaigne, J.-P., & Frohman, L. A. (2006, March). Prolonged stimulation of growth hormone (GH) and insulin-like growth factor I secretion by CJC-1295, a long-acting analog of GH-releasing hormone in healthy adults. *Journal of Clinical Endocrinology and Metabolism* 91:799–805. doi: 10.1210/jc.2005-1536

GH 191

Colao, A., Di Somma, C., Spiezia, S., Rota, F., Pivonello, R., Savastano, S., & Lombardi, G. (2006, June). The natural history of partial growth hormone deficiency in adults: a prospective study on the cardiovascular risk and atherosclerosis. *Journal of Clinical Endocrinology and Metabolism* 91(6): 2191–2200. doi:10.1210/jc.2005-2566

GH Deficiency and Replacement

Luque, R.M., & Kineman, R.D. (2006 June). Impact of obesity on the growth hormone axis: evidence for a direct inhibitory effect of hyperinsulinemia on pituitary function. *Endocrinology 147*(6): 2754–2763. doi:10.1210/en.2005-1549

Molitch, M. E., Clemmons, D. R., Malozowski, S., Merriam, G. R. & Vance, M. L. (2011, June). Evaluation and treatment of adult growth hormone deficiency: an Endocrine Society clinical practice guideline. *Journal of Clinical Endocrinology and Metabolism 96*(6): 1587–1609. doi:10.1210/jc.2011-0179

Pasarica, M., Zachwieja, J. J., Dejonge, L., Redman, S., & Smith, S. R., (2007, November). Effect of growth hormone on body composition and visceral adiposity in middle-aged men with visceral obesity. *Journal of Clinical Endocrinology and Metabolism 92*(11): 4265–4270. doi:10.1210/jc.2007-0786

Rochira V., & Guaraldi, G. (2017, February). Growth hormone deficiency and human immunodeficiency virus. *Best Practices and Research: Clinical Endocrinology & Metabolism 31*(1):91–111. doi: 10.1016/j.beem.2017.02.006

Vahl, N., Jorgensen, J. O., Skjaerbaek, C., Veldhuis, J. D., Orskov, H., & Christiansen, J. S. (1997, June 1). Abdominal adiposity rather than age and sex predicts mass and regularity of GH secretion in healthy adults. *American Journal of Physiology 272* (6 Pt 1): E1108–E1116. doi:10.1152/ajpendo.1997.272.6.E1108

GHRPs

Filigheddu, N., Gnocchi, V. F., Coscia, M., Cappelli, M., Porporato, P. E.,. . . . Graziani, A.(2007, March). Ghrelin and des-acyl ghrelin promote differentiation and fusion of C2C12 skeletal muscle cells. *Molecular Biology of the Cell 18*(3): 986–994. doi: 10.1091/mbc.E06-05-0402

Ghè, C., Cassoni, P., Catapano, F., Marrocco, T., Deghenghi, R., . . . Papotti, M. (2002, February). The antiproliferative effect of synthetic peptidyl GH secretagogues in human CALU-1 lung carcinoma cells. *Endocrinology 143*(2), 484–491. doi:10.1210/endo.143.2.8654

Granado, M., Martín, A. I., López-Menduiña, M., López-Calderón, A., Villanúa, M. A. (2008, January). GH-releasing peptide-2 administration prevents liver inflammatory response in endotoxemia. *American Journal of Physiology: Endocrinology & Metabolism 294*(1): E131–E141. doi:10.1152/ajpendo.00308.2007

Iantorno, M., Chen, H., Kim, J., Tesauro, M., Lauro, D., Cardillo, C. & Quon, M. J. (2007, March 1). Ghrelin has novel vascular actions that mimic PI 3-kinase-dependent actions of insulin to stimulate production of NO from endothelial cells. *American Journal of Physiology 292*(3): E756–E764. doi:10.1152/ajpendo.00570.2006

Ohnuma, T., Ali, M. A., Adigun, R. (2020, June 23) Cancer, anorexia and cachexia. In StatPearls [Internet]. Treasure Island, FL: StatPearls Publishing. Retrieved from https://www.ncbi.nlm.nih.gov/books/NBK430977

Ipamorelin

Beck, D.E., Sweeney, W. B., McCarter, M.D., Ipamorelin 201 Study Group. (2014, October 20). Prospective, randomized, controlled, proof-of-concept study of the Ghrelin mimetic ipamorelin for the management of postoperative ileus in bowel resection patients. *International Journal of Colorectal Disease* *29*(12):1527–34. doi: 10.1007/s00384-014-2030-8

Raun, K., Hansen, B. S., Johansen, N. L., Thøgersen, H., Madsen, K., Ankersen, M., & Andersen, P. H. (1998, November). Ipamorelin, the first selective growth hormone secretagogue. *European Journal of Endocrinology* *139*(5):552–61. doi:10.1530/eje.0.1390552

GHRP-6

Berlanga-Acosta, J., Nieto, G. G., Lopez-Mola, E., & Herrera-Martinez, L. (2016). Growth hormone releasing peptide-6 (GHRP-6) and other related secretagogue synthetic peptides: A mine of medical potentials for unmet medical needs. *Integrative Molecular Medicine 3*: 616–623. doi: 10.15761 /IMM.1000213

Frieboes, R.M., Murck, H., Maier, P., Schier, T., Holsboer, F., & Steiger A. (1995, May). Growth hormone-releasing peptide-6 stimulates sleep, growth hormone, ACTH and cortisol release in normal man. *Neuroendocrinology 61*:584-589. doi:10.1159/000126883

Kelestimur, F., Popovic, V., Leal, A., Van Dam, P. S., Torres, E., . . . Casanueva, F. F. (2006, June). Effect of obesity and morbid obesity on the growth hormone (GH) secretion elicited by the combined GHRH + GHRP-6 test. *Clinical Endocrinology (Oxf.) 64*(6):667–71. doi:10.1111/j.1365-2265.2006.02525.x

Svenson, J., Lall, S., Dickson, S. L., Bengtsson, B. A., Rømer, J., . . . Jansson, J. O., (2000, June). The GH secretagogues ipamorelin and GH-releasing peptide-6 increase bone mineral content in adult female rats. *Journal of Endocrinology 165*(3): 569–77. doi:10.1677/joe.0.1650569

GHRP-2

Laferrère, B., Abraham, C., Russell, C. D., & Bowers, C.Y. (2005, February). Growth hormone releasing peptide-2 (GHRP-2), like ghrelin, increases food intake in healthy men. *Journal of Clinical Endocrinology & Metabolism 90*: 611–614. doi:10.1210/jc.2004-1719

MK-0677

Jacks, T., Smith, R., Judith, F., Schleim, K., Frazier, E., . . . Hickey, G. (1996, December). MK-0677, a potent, novel orally active growth hormone (GH) secretagogue: GH, Insulin-like growth factor 1 and other hormonal responses in beagles. *Endocrinology 137*(12): 5285–89. doi: 10.1210/endo.137 .12.8940347

Meyer, R. M., Burgos-Robles, A., Liu, E., Correia, S. S., & Goosens, K. A. (2014). A ghrelin-growth hormone axis drives stress-induced vulnerability to enhanced fear. *Molecular Psychiatry 19*:1284–94. doi:10.1038/mp.2013.135

Nutrition and GH

Bowers, C.Y., Granda, R., Mohan, S., Kuipers, J., Baylink, D., & Veldhuis, J.D. (2004, May). Sustained elevation of pulsatile growth hormone (GH) secretion and insulin-like growth factor I (IGF-I), IGF-binding protein-3 (IGFBP-3), and IGFBP-5 concentrations during 30-day continuous subcutaneous infusion of GH-releasing peptide-2 in older men and women. *Journal of Clinical Endocrinology & Metabolism 89*: 2290–300. doi: 10.1210 /jc.2003-031799

Epithalon

Anisimov, V. N., Khavinson, V. K., Popovich, I. G., Zabezhinski, M. A., Alimova, I. N., . . . Yashin, A.I. (2003, August). Effect of epitalon on biomarkers of aging, life span and spontaneous tumor incidence in female Swiss-derived SHR mice. *Biogerontology 4*(4):193–202. doi:10.1023/a:1025114230714

Khavinson, V., Diomede, F., Mironova, E., Linkova, N., Trofimova, S., . . . Sinjari, B. (2020, January 30). AEDG Peptide (Epitalon) stimulates gene expression and protein synthesis during neurogenesis: possible epigenetic mechanism. *Molecules 25*(3):609. doi:10.3390/molecules25030609

Khavinson, V. & Morozov, V. G. (2003, June–August). Peptides of pineal gland and thymus prolong human life. *Neuro Endocrinology Letters 24*(3–4): 233–40.

Khavinson, V. K., Bondarev, I. E., & Butyugov, A. A. (2003, June). Epithalon peptide induces telomerase activity and telomere elongation in human somatic cells, *Bulletin of Experimental Biology and Medicine 135*(5): 692–695, doi:10.1023/a:1025493705728

Korkushko, O. V. (1996). The results of 30-month use of thymalin and epithalamin in people with the manifestations of an accelerated aging. International Symposium, Gerontological Aspects of Peptide Regulation of Organism Functions, St. Petersburg: Nauka.

Korkushko, O., Khavinson, V.Kh., Shatilo, V.B., Antonyuk-Shcheglova, I.A. (2006, September). Geroprotective effect of epithalamine (pineal gland peptide preparation) in elderly subjects with accelerated aging. *Bulletin of Experimental Biology & Medicine 142*(3): 356–9. doi: 10.1007/s10517-006-0365-z

Korkushko, O. V. (1996). The results of 30-month use of thymalin and epithalamin in people with the manifestations of an accelerated aging. International Symposium, Gerontological Aspects of Peptide Regulation of Organism Functions, St. Petersburg: Nauka.

Anti-Aging

Korkushko, O. V. (1996). The results of 30-month use of thymalin and epithalamin in people with the manifestations of an accelerated aging. International Symposium, Gerontological Aspects of Peptide Regulation of Organism Functions, St. Petersburg: Nauka.

Korkusko, O. V. (2002, November 8-11). Peptide preparations from the thymus and pineal gland may be used for the slowing down of accelerated human aging. Proceedings of the third European Congress of Biogerontology. Florence, Italy. *Biogerontology 3*(1): 65.

Peptide regulation of Aging. (2001, July 1–6). Proceedings of the 17th World Congress of the International Association of Gerontology, Vancouver, Canada. *Gerontology 47*(1): 545.

Anti-Tumor

Khavinson, V. Kh. (1997, May 28–31). Bioregulating therapy as a new direction in Medicine. Proceedings of the National Conference of Gerontology and Geriatrics. Bucharest, Romainia. 205.

CHAPTER 5

Immune Support

Aristizábal, B. & González, A. (2013, July 18) Chapter 2: Innate immune system. In Anaya, J.M., Shoenfeld, Y., Rojas-Villarraga, A, et al., editors. *Autoimmunity: From Bench to Bedside* [Internet]. Bogota, Colombia: El Rosario University Press. Retrieved from https://www.ncbi.nlm.nih.gov/books/NBK459455/

Bernstein, K. E., Khan, Z., Giani, J. F., Cao, D.-Y., Bernstein, E. A., & Shen, X. Z. (2018, May). Angiotensin-converting enzyme in innate and adaptive immunity. *Nature Reviews Nephrology, 14*(5), 325–336. doi: 10.1038/nrneph.2018.15

Carvajal, G., Rodríguez-Vita, J., Rodrigues-Díez, R., Sánchez-López, E., Rupérez, M., Cartier, C., . . . Ruiz-Ortega, M. (2008, September). Angiotensin II activates the Smad pathway during epithelial mesenchymal transdifferentiation. *Kidney International, 74*(5), 585–595. doi:10.1038/ki.2008.213

Gurwitz, D. (2020, March 4). Angiotensin receptor blockers as tentative SARS-CoV-2 therapeutics. *Drug Development Research.* doi:10.1002/ddr.21656

Huss, D.J., Winger, R.C., Cox, G.M., Guerau-de-Arellano, M., Yang, Y., Racke, M. K., Lovett-Racke, A. E. (2011, October). TGF-β signaling via Smad4 drives IL-10 production in effector Th1 cells and reduces T-cell trafficking in EAE [published correction appears in *European Journal of Immunology* in November 2012; *42*(11):3084]. *European Journal of Immunology 41*(10): 2987–2996. doi:10.1002/eji.201141666

Komai, T., Inoue, M., Okamura, T., Morita, K., Iwasaki, Y., . . . Fujio, K. (2018 June 14). Transforming growth factor-β and interleukin-10 synergistically regulate humoral immunity via modulating metabolic signals. *Frontiers in Immunology 9*:1364. doi:10.3389/fimmu.2018.01364

Li, W., Moore, M. J., Vasilieva, N., Sui, J., Wong, S. K., Berne, M. A., . . . Farzan, M. (2003, November 27). Angiotensin-converting enzyme 2 is a functional receptor for the SARS coronavirus. *Nature, 426*(6965), 450–454. doi:10.1038/nature02145

Passos-Silva, D. G., Brandan, E., & Santos, R. A. S. (2015, April 3). Angiotensins as therapeutic targets beyond heart disease. *Trends in Pharmacological Sciences, 36*(5), 310–320. doi:10.1016/j.tips.2015.03.001

Steinman, R. M. & Hemmi, H. (2006). Dendritic cells: translating innate to adaptive immunity. In Pulendran, B., & Ahmed, R. (eds) *From Innate Immunity to Immunological Memory. Current Topics in Microbiology and Immunology, vol 311.* Berlin, Heidelberg: Springer. 17–58. Retrieved from https://link.springer.com/chapter/10.1007/3-540-32636-7_2

Sungnak, W., Huang, N., Bécavin, C., Berg, M., Queen, R., . . . Barnes, J. L. (2020, May). SARS-CoV-2 entry factors are highly expressed in nasal epithelial cells together with innate immune genes. *Nature Medicine 26*(5): 681–687. doi:10.1038/s41591-020-0868-6

Wang, L., Zhu, Q., Lu, A., Liu, X., Zhang, L., Xu, C., . . . Yang, T. (2017). Sodium butyrate suppresses angiotensin II-induced hypertension by inhibition of renal (pro)renin receptor and intrarenal renin–angiotensin system. *Journal of Hypertension, 35*(9), 1899–1908. doi: 10.1097/HJH.0000000000001378

CHAPTER 6

MGF

Geusens, P. P. & Boonen S. (2002). Osteoporosis and the growth hormone-insulin like growth factor axis. *Hormone Research 58*(Supp 3): 49–55. doi: 10.1159/000066483

Goldspink ,G. (2004, June). Age-related muscle loss and progressive dysfunction in mechanosensitive growth factor signaling. *Annals of the New York Academy of Sciences 1019*: 294–8. doi: 10.1196/annals.1297.050

Hamid, M., Orrell, R. W., Cobbold, M., Goldspink, G. & Harridge, S. D. R. (2003, February 15). Expression of IGF-1 splice variants in young and old human skeletal muscle after high resistance exercise. *Journal of Physiology 547*(1): 247–54. doi:10.1113/jphysiol.2002.032136

Kandalla, P.K., Goldspink, G., Butler-Browne, G., Mouly, V. (2011, April). Mechano Growth Factor E peptide (MGF-E), derived from an isoform of the IGF-1, activates human muscle progenitor cells and induces an increase in their fusion potential at different ages. *Mechanics of Ageing and Development 132*(4):154–162. doi:10.1016/j.mad.2011.02.007.

Lalancette-Hebert, M., Gowing, G., Simard, A., Weng, Y. C., Kriz, J. (2007, March 7). Selective ablation of proliferating microglia cells exacerbates ischemic injury in the brain. *Journal of Neuroscience 27*(10):2596–2605. doi:10.1523/JNEUROSCI.5360-06.2007

Owino, V., Yang, S. Y., Goldspink, G. (2001, September 14). Age-related loss of skeletal muscle function and the inability to express the autocrine form of insulin-like growth factor-1 (MGF) in response to mechanical overload. *FEBS Letters 505*(2): 259–263. doi:10.1016/s0014-5793(01)02825-3

Philippou, A., Papageorgiou, E., Bogdanis, G., Halapas, A., Sourla, A., . . . Koutsilieris, M. (2009, July–August) Expression of IGF-1 isoforms after exercise induced muscle damage in humans: characterization of the MGF E peptide action in vitro. *In Vivo 23*(4):567–75.

Yang, S., Alnaqeeb, M., Simpson, H., & Goldspink, G. (1996, August). Cloning and characterization of an IGF-I isoform expressed in skeletal muscle subjected to stretch. *Journal of Muscle Research & Cell Motility 17*: 487–95. doi:10.1007/BF00123364

IGF-1Ec: MGF and Heart

Carpenter, V., Matthews, K., Devlin, G., Stuart. S., Jensen, J., . . . McMahon, C. (2008, February). Mechano-growth factor reduces loss of cardiac function in acute myocardial infarction, *Heart, Lung and Circulation 17*:33–39. doi:10.1016/j.hlc.2007.04.013

Mavrommatis, E., Shioura, K. M., Los, T., & Goldspink, P. (2013). The E-domain region of mechano-growth factor inhibits cellular apoptosis and preserves cardiac function during myocardial infarction. *Molecular and Cellular Biochemistry 38*(1–2):69–83. doi: 10.1007/s11010-013-1689-4

Stavropoulou, A., Halapas, A., Sourla, A., Philippou, A., Papageorgiou, E., Papalois, A., & Koutsilieris, M. (2009, May–June). IGF-1 expression in infarcted myocardium and MGF E peptide actions in rat cardiomyocytes in vitro. *Molecular Medicine 15*:127–135. doi: 10.2119/molmed.2009.00012

IGF-1Ec and the Brain

Carro, E., Trejo, J. L., Nunez, A., & Torres-Aleman, I. (2003, April). Brain repair and neuroprotection by serum insulin-like growth factor I. *Molecular Neurobiology 27*(2):153–162. doi:10.1385/MN:27:2:153

Dluzniewska, J., Sarnowska, A., Beresewicz, M., Johnson, I., Srai, S. K. S., . . . Zabłocka, B. (2005, November). A strong neuroprotective effect of the autonomous C-terminal peptide of IGF-1 Ec (MGF) in brain ischemia. *FASEB Journal 19*(13):1896–1898. doi:10.1096/fj.05-3786fje

Doroudian, G., Pinney, J., Ayala, P., Los, T., Desai, T.A., & Russell, B. (2014, October). Sustained delivery of MGF peptide from microrods attracts stem cells and reduces apoptosis of myocytes. *Biomedical Microdevices 16*(5):705–15. doi:10.1007/s10544-014-9875-z

Geusens, P. P. & Boonen S. (2002) Osteoporosis and the growth hormone-insulin like growth factor axis. *Hormone Research 58*(Supp 3): 49–55. doi: 10.1159/000066483

Lalancette-Hebert, M., Gowing, G., Simard, A., Weng, Y.C., Kriz, J. (2007 March 7). Selective ablation of proliferating microglia cells exacerbates ischemic injury in the brain. *Journal of Neuroscience 27*(10):2596–2605. doi:10.1523/JNEUROSCI.5360-06.2007

IGF-1

Chen, W., Wang, S., Tian, T., Bai, J., Hu, Z., . . . Shen., H. (2009, December). Phenotypes and genotypes of insulin-like growth factor 1, IGF-binding protein-3 and cancer risk: evidence from 96 studies. *European Journal of Human Genetics 17*(12): 1668–1675. doi:10.1038/ejhg.2009.86

Geusens, P. P. & Boonen S. (2002) Osteoporosis and the growth hormone-insulin like growth factor axis. *Hormone Research 58*(Supp 3): 49–55. doi: 10.1159/000066483

Junilla, R. K., List, E. O., Berryman, D .E., Murrey, J. W., & Kopchick J. J. (2013, June). The GH/IGF-1 axis in aging and longevity. *Nature Reviews Endocrinology 9*(6):366–76. doi: 10.1038/nrendo.2013.67

CHAPTER 7

Cell Repair

Dierickx, P., Van Laake, L. W., & Geijsen, N. (2018) Circadian clocks: from stem cells to tissue homeostasis and regeneration. *EMBO Reports 19*(1): 18–28. doi:10.15252/embr.201745130

Hoyle, N. P., Seinkmane, E., Putker, M., Feeney, K. A., Krogager, T. P., Chesham, J. E., … O'Neill, J. S. (2017). Circadian actin dynamics drive rhythmic fibroblast mobilization during wound healing. *Science Translational Medicine* 9(415): eaal2774. doi: 10.1126/scitranslmed.aal2774

Rogers, E.H., Hunt, J.A., Pekovic-Vaughan, V. (2018, December). Adult stem cell maintenance and tissue regeneration around the clock: do impaired stem cell clocks drive age-associated tissue degeneration? *Biogerontology* 19(6):497–517. doi: 10.1007/s10522-018-9772-6

Serin, Y., & Acar Tek, N. (2019). Effect of circadian rhythm on metabolic processes and the regulation of energy balance. *Annals of Nutrition and Metabolism 74, 322–330.* doi:10.1159/000500071

Zhang, R., Lahens, N.F., Ballance, H.I., Hughes, M.E., & Hogenesch, J.B. (2014, November 11). A circadian gene expression atlas in mammals: implications for biology and medicine. *Proceedings of the National Academy of Science of the United States of America 111*(45):16219–16224. doi:10.1073/pnas.1408886111

GHK-CU

Bishop, J.B., Phillips, L.G., Mustoe, T.A., VanderZee, A.J., Wiersema, L., … Roach, Robson, M.C. (1992, August). A prospective randomized evaluator-blinded trial of two potential wound healing agents for the treatment of venous stasis ulcers. *Journal of Vascular Surgery 16*(2): 251–257. doi:10.1067/mva.1992.37086

Lamb, J (2007, January 1). The Connectivity Map: a new tool for biomedical research. *Nature Reviews Cancer. 7*(1): 54–60. doi:10.1038/nrc2044

Pickart, L. (2008) The human tri-peptide GHK and tissue remodeling. *Journal of Biomaterials Science: Polymer Edition 19*(8):969–988. doi:10.1163/156856208784909435

IGF-1

Chen, W., Wang, S., Tian, T., Bai, J., Hu, Z., … Shen, H. (2009, December). Phenotypes and genotypes of insulin-like growth factor 1, IGF-binding protein-3 and cancer risk: evidence from 96 studies. *European Journal of Human Genetics 17*(12): 1668–1675. doi:10.1038/ejhg.2009.86

Geusens, P. P. & Boonen S. (2002). Osteoporosis and the growth hormone-insulin like growth factor axis. *Hormone Research 58*(Supp 3): 49–55. doi: 10.1159/000066483

Junilla, R. K., List, E. O., Berryman, D. E., Murrey, J. W., &. Kopchick, J. J. (2013, June) The GH/IGF-1 axis in aging and longevity. *Nature Reviews Endocrinology 9*(6):366–76. doi:10.1038/nrendo.2013.67

Lalancette-Hebert, M., Gowing, G., Simard, A., Weng, Y.C., Kriz, J. (2007, March 7). Selective ablation of proliferating microglia cells exacerbates ischemic injury in the brain. *Journal of Neuroscience 27*(10):2596–2605. doi:10.1523/JNEUROSCI.5360-06.2007

TB4 (Thymosin Beta 4)

Badamchian, M., Fagarasan, M. O., Danner, R. L., Suffredini, A. F., Dama-vandy, H., & Goldstein, A.L. (2003, August) Thymosin beta 4 reduces lethal-ity and down-regulates inflammatory mediators in endotoxin-induced septic shock. *International Immunopharmacology 3*(8):1225–33. doi:10.1016/S1567-5769(03)00024-9

Goldstein, A.L., Hannappei, E., Sosne, G., & Kleinman, H. K. (2012, January) Thymosin beta4: a multifunctional regenerative peptide: Basic properties and clinical applications. *Expert Opinion on Biological Therapy 12*(1):37–51. doi:10.1517/14712598.2012.634793

Grant, D.S., Rose, W., Yaen, C., Goldstein, A., Martinez, J., & Kleinman, H. (1999). Thymosin Beta 4 enhances endothelial cell differentiation and angio-genesis. *Angiogenesis 3*(2):125–35. doi:10.1023/a:1009041911493

Jiang, Y., Han, T., Zhang, Z., Li, M., Qi, F.-X., Zhang, Y., & Ji, Y.-L. (2017, July 8). Potential role of thymosin beta 4 in the treatment of nonalcoholic fatty liver disease. *Chronic Diseases and Translational Medicine 3*(3):165–68. doi:10.1016/j.cdtm.2017.06.003

Kleinman, H.K., Sosne, G. (2016, May 23). Thymosin β4 promotes dermal heal-ing. *Vitamins and Hormones 102*:251–75. doi:10.1016/bs.vh.2016.04.005

Malinda, K. M., Sidhu, G. S., Mani, H., Banaudha, K., Maheshwari, R. K., Goldstein, A.L., & Kleinman, H.K. (1999, September). Thymosin beta4 accelerates wound healing. *Journal of Investigative Dermatology 113*(3):364–8. doi:10.1046/j.1523-1747.1999.00708.x

Philp, D., Goldstein, A.L., Kleinman, H.K. (2004, February). Thymosin Beta 4 promotes angiogenesis, wound healing, and hair follicle development. *Mecha-nisms of Ageing and Development 125*(2):113–5. doi: 10.1016/j.mad.2003.11.005

Popoli, P.R., Pepponi, A., Martire, A., Armida, M., Pèzzola, A., . . . Garaci, E. (2007, October 10). Neuroprotective Effects of Thymosin β4 in Experimental Models of Excitotoxicity. *Annals of the New York Academy of Sciences 1112*: 219–224.doi: 10.1196/annals.1415.033

Reti, R., Kwon, E., Qui, P., Wheater, M., & Sosne, G. (2008, October). Thymosin β4 is cytoprotective in human gingival fibroblasts. *European Journal of Oral Sciences 116*(5):424–30. doi:10.1111/j.1600-0722.2008.00569.x

Sanders, M. C., Goldstein, A. L., Wang, Y. (1992, April 30). Thymosin β4 (Fx peptide) is a potent regulator of actin polymerization in living cells. *Proceedings of the National Academy of Sciences of the United States of America 89*(10):4678–4682. doi:10.1073/pnas.89.10.4678

Shrivastava, S., Srivastava, D., Olson, E. N., DiMaio, M., & Bock-Marquette, I. (2010). Thymosin beta4 and cardiac repair. *Annal of the New York Academy of Sciences 1194*: 87–96. doi: 10.1111/j.1749-6632.2010.05468.x

Sosne, G., Rimmer, D., Kleinman, H.K., Ousler, G. (2016, July 5). Thymosin Beta 4: A Potential Novel Therapy for Neurotrophic Keratopathy, Dry Eye, and Ocular Surface Diseases. *Vitamins & Hormones 102*: 277–306. doi: 10.1016/bs.vh.2016.04.012

Yarmola, E.G., Kilmenko, E.S., Fujita, G., Bubb, M. (2007, October 7). Thymosin β4: actin regulation and more. *Annals of the New York Academy of Sciences 1112*(1):76–85. doi: 10.1196/annals.1415.008

BPC 157

Duzel, A. Vlainic, J., Antunovic, M., Malekinusic, D., Vrdoljak, B., . . . Sikiric, P. (2017, December 28). Stable gastric pentadecapeptide BPC 157 in the treatment of colitis and ischemia and reperfusion in rats: new insights. *World Journal of Gastroenterology 23*(48): 8465–88. doi: 10.3748/wjg.v23.i48.8465

Grgic, T., Grgic, D., Drmic, D., Sever, A. Z., Petrovic, I., . . . Sikiric, P. (2016, June 5). Stable gastric pentadecapeptide BPC 157 heals rat colovesical fistula. *European Journal of Pharmacology 780*:1–7. doi: 10.1016/j.ejphar.2016.02.038

Pevec, D., Novinscak, T., Brcic, L., Sipos, K., Jukic, I., . . . Sikiric, P. (2010, March). Impact of pentadecapeptide BPC 157 on muscle healing impaired by systemic corticosteroid application. *Medical Science Monitor 16*(3): BR81–88.

Seiwerth S., Brcic, L., Vuletic, L. B., Kolenc, D., Aralica, G., . . . Sikiric P. (2014). BPC157 and blood vessels. *Current Pharmaceutical Designs 20*(7):1121–35. doi:10.2174/13816128113199990421

Sikiric P., Seiwerth, S., Rucman, R., Kolenc, D., Vuletic, L. B., . . . Vlainic, J. (2016, November). Brain-gut axis and pentadecapeptide BPC 157: Theoretical and Practical Implications. *Current Neuropharmacology 14*(8): 957–865. doi: 10.2174/1570159X13666160502153022

Sikiric, P., Seiwerth, S., Rucman, R., Turkovic, B., Rokotov, D. S., . . . Sebecic, B. (2012). Focus on ulcerative colitis: stable gastric pentadecapeptide BPC 157. *Current Medical Chemistry 19*(1): 126–32. doi: 10.2174/092986712803414015

Strinic, D., Halle, Z. B., Luetic, K., Nedic, A., Petrovic, I., . . . Sikiric P. (2017, October 1). BPC 157 counteracts QTc prolongatino induced by haloperidol, fluphenazine, clozapine, olanzapine, quetiapine, sulpiride and metoclopramide in rats. *Life Sciences 186*(1):66–79. doi: 10.1016/j.lfs.2017.08.006

CHAPTER 8

Immune Upregulation

Thymosin Alpha 1 (TA1)

An, T. T., Liu, X.-Y., Fang, J. & Wu, M.-N. (2004, November). [Primary assessment of treatment effect of thymosin alpha1 on chemotherapy-induced neurotoxicity][Article in Chinese]. *Ai Zheng 23*(Suppl 11):1428–30.

Ershler, W.B., Gravenstein, S., & Geloo, Z. S. (2007, October 10). Thymosin alpha 1 as an adjunct to influenza vaccine in the elderly: rational and trial summaries. *Annals of the New York Academy of Sciences 1112*(1):375–84. doi:10.1196/annals.1415.050

Giuliani, C., Napolitano, G., Mastino, A., Di Vincenzo, S., D'Agostini, C., . . . Favalli, C. (2000, March). Thymosin-alpha1 regulates MHC class I expression in FRTL-5 cells at transcriptional level. *European Journal of Immunology 30*(3): 778–86. doi:10.1002/1521-4141(200003)30:3<778::AID-IMMU778>3.0.CO;2-I

Goldstein, A.L. (2007, September) History of the discovery of the thymosins. *Annals of New York Academy of Sciences 1112*: 1–13. doi: 10.1196/annals.1415.045

Liu, D., Wang, H.-M., Wang, T., Zhang, Y.-M., & Zhu, X. (2016, September). The efficacy of thymosin alpha 1 as immunomodulatory treatment for sepsis: a systematic review of randomized controlled trials. *BMC Infectious Diseases 16*: 488. doi:10.1186/s12879-016-1823-5

Matteucci, C., Grelli, S., Balestrieri, E., Minutolo, A., Argaw-Denboba, A., . . . Garaci, E. (2017, February). Thymosin alpha 1 and HIV-1: recent advances and future perspectives. *Future Microbiology 12*:141–155. doi: 10.2217/fmb-2016-0125

Rasi, G., Terzoli, E., Izzo, F., Pierimarchi, P., Ranuzzi, M., . . . Garaci, E. (2000, April). Combined treatment with thymosin alpha 1 and low-dose interferon alpha after dacarbazine in advanced melanoma. *Melanoma Research 10*(2):189–192.

Romani, L., Moretti, S., Fallarino, F., Bozza, S., Ruggeri, L., . . . Garaci, E. (2012, October). Jack of all trades: thymosin alpha-1 and its pleiotropy. *Annals of the New York Academy of Sciences 1269*:1–6. doi:10.1111/j.1749-6632.2012.06716.x

Yang, X., Qian, F., He, H., Liu, K.-J., Lan, Y.-Z., . . . Wu, Y.-Z. (2012). Effect of thymosin alpha-1 on subpopulations of Th1, Th2, Th17 and regulatory T cells (Tregs) in vitro. *Brazilian Journal of Medicine and Biological Research 45*(1):25–32. doi:10.1590/S0100-879X2011007500159

B4

Bollini, S., Riley, P. R., & Smart, N. (2015, June 22). Thymosin β4: multiple functions in protection, repair and regeneration of the mammalian heart. *Expert Opinion on Biological Therapy 15*(Suppl 1): S163–74. doi: 10.1517/14712598.2015.1022526

Goldstein, A.L., & Kleinman, H. K. (2015, June 22). Advances in the basic and clinical applications of Thymosin Beta 4. *Expert Opinion on Biological Therapy 15*(Suppl 1): S139–45. doi:10.1517/14712598.2015.1011617

Role of the Melanocortin System

Adan, R. A. H., Tisesjema, B., Hillebrand, J. J. G., la Fleur, S. E., Kas, M. J. H., & de Krom, M. (2006, December). The MC4 receptor and control of appetite. *British Journal of Pharmacology 149*(7):815–27. doi:10.1038/sj.bjp.0706929

DeSilva, A., do Carmo, J. M., Wang, Z., & Hall, J. E. (2014, May). The brain melanocortin system, sympathetic control and obesity hypertension. *Physiology (Bethesda). 29*(3): 196–202. doi:10.1152/physiol.00061.2013

Guiliani, D., Ottani, A., Altavilla, D., Bazzani, C., Squadrito, F., & Guarin, S. (2010). Melanocortins and the cholinergic anti-inflammatory pathway. In Catania A. (eds), *Melanocortins: Multiple Actions and Therapeutic Potential. Advances in Experimental Medicine and Biology*, vol 681. New York: Springer, 71–87.

Pavlov, V.A., & Tracey, K. J. (2006, December). Controlling inflammation: the cholinergic anti-inflammatory pathway. *Biochemical Society Transactions 34*(6):1037–40. doi:10.1042/BST0341037

Schwartz, M. W., Seeley, R. J., Woods, S. C., Weigle, D. S., Campfield, L. A., Burn, P., & Baskin, D. G. (1997, December). Leptin increases hypothalamic pro-opiomelanocortin mRNA expression in the rostral arcuate nucleus. *Diabetes 46*: 2119-2123. doi:10.2337/diab.46.12.2119

Wikberg, J. E. (2001). Melanocortin receptors: new opportunities in drug discovery. *Expert Opinion on Therapeutic Patents 11*(1): 61–76. doi: 10.1517/13543776.11.1.61

Melanotan

Dorr, R.T., Lines, R., Levine, N., Brooks, C., Xiang, L., Hruby, V. J., & Hadley, M. E. (1996, April 12). Evaluation of melanotan II a superpotent cyclic melanotropic peptide in a pilot phase-1 clinical study. *Life Sciences 58*(20):1777–84. doi:10.1016/0024-3205(96)00160-9

Kim, E.S., Garnock-Jones, K. P. (2016, March 15) Afamelanotide: A review in erythropoietic protoporphyria. *American Journal of Clinical Dermatology 7*:179–185. doi:10.1007/s40257-016-0184-6

Langendock, J. G., Balwani, M., Anderson, K. E., Bonkovsky, H. L., Anstey, A. V., . . . Desnick, R. J. (2015, July 2). Afamelanotide for Erythropoietic Protoporphyria. *New England Journal of Medicine 373*(1):48–59. doi:10.1056 /NEJMoa1411481

Mykicki, N., Herrmann, A.M., Schwab, N., Deenen, R., Sparwasser, T., . . . Loser, K. (2016, October 26). Melanocortin-1 receptor activation is neuroprotective in mouse models of neuroinflammatory disease. *Science Translational Medicine 8*(362):362ra146. doi: 10.1126/scitranslmed.aaf8732

Melanotan II

Breindahl, T., Evans-Brown, M., Hindersson, P., McVeigh, J., Bellis, M., Stensballe, A., & Kimergård, A. (2015, February). Identification and characterization by LC-UV-MS/MS of melanotan II skin-tanning products sold illegally on the Internet. *Drug Testing and Analysis 7*(2): 164–172. doi:10.1002 /dta.1655

Devlin, J., Pomerleau, A., Foote, J. (2013, May). Melanotan II overdose associated with priapism. *Clinical Toxicology (Phila) 51*(4): 383. doi:10.3109/15563650 .2013.784775

Dorr, R.T., Lines, R., Levine, N., Brooks, C., Xiang, L., Hruby, V. J., & Hadley, M. E. (1996, April 12) Evaluation of melanotan II a superpotent cyclic melanotropic peptide in a pilot phase-1 clinical study. *Life Sciences 58*(20):1777–84. doi:10.1016/0024-3205(96)00160-9

Loram, L.C., Culp, M.E., Connolly-Strong, E.C., & Sturgill-Koszycki, S. (2015, February). Melanocortin peptides: potential targets in systemic lupus erythematosus. *Inflammation 38*(1): 260–71. doi:10.1007/s10753-014-0029-5

Nelson, M.E., Bryant, S.M., & Aks, S.E. (2012, December). Melanotan II injection resulting in systemic toxicity and rhabdomyolysis. *Clinical Toxicology (Phila.) 50*(10):1169–73. doi: 10.3109/15563650.2012.740637

Olney, J.J., Navarro M., Thiele T.E. (2014, June 3). Targeting central melanocortin receptors: a promising novel approach for treating alcohol abuse disorders. *Frontiers in Neuroscience 8*:128. doi: 10.3389/fnins.2014.00128

Taylor, A.W., Yee, D.G., Nishida, T., & Namba, K. (2000). Neuropeptide regulation of immunity. The immunosuppressive activity of alpha-melanocyte stimulating hormone (alpha-MSH). *Annals of the New York Academy of Sciences 917*: 239–47. doi:10.1111/j.1749-6632.2000.tb05389.x

Wessells, H., Levine, N., Hadley, M.E., Dorr, R., & Hruby, V. (2000, October). Melanocortin receptor agonists, penile erection, and sexual motivation: human studies with Melanotan II. *International Journal of Impotence Research 12*(Suppl 4): S74–9. doi: 10.1038/sj.ijir.3900582

Sleep and DSP

Afaghi, A., O'Connor, H., & Chow, C. (2008, August). Acute Effects of the Very Low Carbohydrate Diet on Sleep Indices. *Nutritional Neuroscience 11*(4), 146–154. doi:10.1179/147683008X301540

Graf, M. V., Hunter, C.A., Kastin, A.J. (1984, July). Presence of delta-sleep inducing peptide-like material in human milk. *Journal of Clinical Endocrinology & Metabolism 59*(1): 127–132. doi: 10.1210/jcem-59-1-127

Graf, M. V., Kastin, A.J., Coy, D.H., Fischman, A.J. (1985, October). Delta-sleep-inducing peptide reduces CRF-induced corticosterone release. *Neuroendocrinology 41*(4): 353–356. doi:10.1159/000124200

Maquet, P., Degueldre, C., Delfiore, G., Aerts, J., Peters, J. M., Luxen, A., & Franck, G. (1997, April 15). Functional neuroanatomy of human slow wave sleep. *Journal of Neuroscience 17*(8), 2807–2812. doi:10.1523/JNEUROSCI .17-08-02807.1997

Schneider-Helmert, D., Schoenenberger, G.A. (1983). Effects of DSIP in man. Multifunctional psychophysiological properties besides induction of natural sleep. *Neuropsychobiology 9*(4): 197–206. doi:10.1159/000117964

CHAPTER 9

Metabolism and Weight Loss

Heffernan, M.A., Jiang, W. J., Thorburn, A.W., Ng, F.M. (2000, September). Effects of oral administration of a synthetic fragment of human growth hormone on lipid metabolism. *American Journal of Physiology: Endocrinology & Metabolism 279*(3): E501–E507. doi:10.1152/ajpendo.2000.279.3.E501

Heffernan, M. A., Thorburn, A. W., Fam, B., Summers, R., Conway-Campbell, B., Waters, M. J., & Ng, F.M. (2001, October 23). Increase of fat oxidation and weight loss in obese mice caused by chronic treatment with human growth hormone or a modified C-terminal fragment. *International Journal of Obesity 25*(10):1442–1449. doi:10.1038/sj.ijo.0801740

Kim, S., Miller, B., Mehta, H. H., Xiao, J., Wan, J., … Cohen, P. (2019, July). The mitochondrial-derived peptide MOTS-c is a regulator of plasma metabolites and enhances insulin sensitivity. *Physiological Reports, 7*(13). doi:10.14814 /phy2.14171

Kwon, D. R., & Park, G. Y. (2015, Summer). Effect of Intra-articular Injection of AOD9604 with or without Hyaluronic Acid in Rabbit Osteoarthritis Model. *Annals of Clinical & Laboratory Science 45* (4), 426–32.

LeBlanc, E. L., Patnode, C. D., Webber, E. M., Redmond, N., Rushkin, M., O'Connor, E. A. (2018, September 18). Behavioral and pharmacotherapy weight loss interventions to prevent obesity-related morbidity and mortality in adults: an updated systematic review for the U.S. Preventive Services Task Force. *JAMA 320*(11): 1172–1191. doi:10.1001/jama.2018.7777

Lu, H., Wei, M., Zhai, Y., Li, Q., Ye, Z.,. . . Lu, Z. (2019, April). MOTS-c peptide regulates adipose homeostasis to prevent ovariectomy-induced metabolic dysfunction. *Journal of Molecular Medicine (Berlin, Germany) 97*(4):473–485. doi:10.1007/s00109-018-01738-w

Mehta, A., Marso, S. P., Neeland, I. J. (2017, March). Liraglutide for weight management: a critical review of the evidence. *Obesity Science & Practice 3*(1): 3–14. doi:10.1002/osp4.84

Ribeiro de Oliveira Longo Schweizer, J., Ribeiro-Oliveira Jr., A., & Bidlingmaier, M. (2018, August). Growth hormone: isoforms, clinical aspects and assays interference. *Clinical Diabetes and Endocrinology 4*:18. doi:10.1186/s40842-018-0068-1

Shah, M., & Vella, A. (2014, September). Effects of GLP-1 on appetite and weight. *Reviews in Endocrine and Metabolic Disorders 15*(3): 181–187. doi:10.1007/s11154-014-9289-5

Sharma, R. (2018, April). Growth Hormone Therapy and Lipid Profile. *Indian Journal of Pediatrics 85*(4): 253–254. doi:10.1007/s12098-018-2638-8

CHAPTER 10

Cerebrolysin

Boado, R.J. (2001, August). Amplification of blood-brain barrier GLUT1 glucose transporter gene expression by brain-derived peptides. *Neuroscience Research 40*(4): 337–42. doi:10.1016/s0168-0102(01)00246-2

Cui, S., Chen, N., Yang, M., Guo, J., Zhou, M., Zhu, C., & He, L. (2019, November 11). Cerebrolysin for vascular dementia. *Cochrane Database of Systematic Reviews*. doi: 10.1002/14651858.CD008900.pub3

Muresanu, D.F., Alvarez, X.A., Moessler, H., Novak, P. H., Stan, A., . . . Popescu, B. O. (2010, December). Persistence of the effects of Cerebrolysin on cognition and qEEG slowing in vascular dementia patients: results of a 3-month extension study. *Journal of Neurological Sciences. 299*(1–2):179–183. doi:10.1016/j.jns.2010.08.040

Rockenstein, E., Adame, A., Mante, M., Larrea, G., Crews, L., Windisch, M., Moessler, H., Masliah, E. (2005, February). Amelioration of the cerebro-vascular amyloidosis in a transgenic model of Alzheimer's disease with the neurotrophic compound Cerebrolysin. *Journal of Neural Transmission (Vienna, Austria) 112*(2): 269–282. doi: 10.1007/s00702-004-0181-4

Rüther, E., Ritter, R., Apecechea, M., Freytag, S., & Windisch, M. (1994, January). Efficacy of the peptidergic nootropic drug Cerebrolysin in patients with senile dementia of the Alzheimer type (SDAT). *Pharmacopsychiatry 27*(1):32–40. doi:10.1055/s-2007-1014271

Wei, Z. H., He, Q. B., Wang, H., Su, B. H., & Chen, H. Z. (2007, February 23). Meta-analysis: the efficacy of nootropic agent Cerebrolysin in the treatment of Alzheimer's disease. *Journal of Neural Transmission 114*(5):629–34. doi:10.1007/s00702-007-0630-y

Semax

Ashmarin, I. P. (2007, September). Glyprolines in regulatory tripeptides. *Neurochemical Journal 1*, 173. doi:10.1134/S1819712407030014

Dolotov, O. V., Karpenko, E. A., Inozemtseva, L. S., Seredenina, T. S., Levitskaya, N. G., . . . Engele, J. (2006, October 30). Semax, an analog of ACTH(4-10) with cognitive effects, regulates BDNF and trkB expression in the rat hippocampus. *Brain Research 1117*(1): 54–60. doi:10.1016/j.brainres.2006.07.108

Inozemtsev, A. N., Bokieva, S. B., Karpukhina, O. V., Gumargalieva, K. Z., Kamensky, A. A., & Myasoedov, N. F. (2016, July 14). Semax prevents learning and memory inhibition by heavy metals. *Doklady Biological Sciences 468*: 112–114. doi:10.1134/S0012496616030066

Storozhevykh, T. P., Tukhbatova, G. R., Senilova, Y. E., Pinelis, V.G., Andreeva, L. A., & Myasoyedov, N.F. (2007, May). Effects of Semax and its Pro-Gly-Pro fragment on calcium homeostasis of neurons and their survival under conditions of glutamate toxicity. *Bulletin of Experimental Biology & Medicine 143*(5):601–604. doi:10.1007/s10517-007-0192-x

Selank

Kolomin, T.A., Iu Agapova, T., Agniullin, Ia V., Shram, S. I., Shadrina, M., . . . Miasoedov, I. F. (2013, May-June).[Transcriptome alteration in hippocampus under the treatment of tuftsin analog Selank] [In Russian]. *Zh Vyssh Nerv Deiat Im I P Pavlova. 63*(3):365–74. doi: 10.7868/s0044467713030052

Meshavkin, V. K., Kost, N. V., Sokolov, O. Y., Andreeva, L. A., & Myasoedov, N.F. (2013, May 8). The influence of oligopeptides on spontaneous carcinogenesis in mice. *Doklady Biological Sciences 449*: 79–81. doi:10.1134/S0012496613020087

Mjasoedov, N. F., Andreeva, L. A., Grigorjeva, M. E., Obergan, T. Y., Shubina, T. A., & Lyapina, L.A. (2014, November 5). The influence of Selank on the parameters of the hemostasis system, lipid profile, and blood sugar level in the course of experimental metabolic syndrome. *Doklady Biological Sciences 458*: 267–270. doi:10.1134/S0012496614050020

Pavlov, T.S., Samonina, G. E., Bakaeva, Z. V., Zolotarev, Yu A., & Guseva A. A. (2007, January). [Selank and Its Metabolites Maintain Homeostasis in the Gastric Mucosa]. Translated from *Byulleten' Eksperimental'noi Biologii i Meditsiny, 143*(1): 51–53. doi: 10.1007/s10517-007-0014-1

Phan, R.T., Saito, M., Basso, K., Niu, H., & Dalla-Favera, R.(2005, October). BCL6 interacts with the transcription factor Miz-1 to suppress the cyclin-dependent kinase inhibitor p21 and cell cycle arrest in germinal center B cells. *Nature Immunology 6*(10):1054–60. doi: 10.1038/ni1245

Seredenin, S. B., Kozlovskaia, M. M., Blednov, Iu A., Kozlovskiĭ, I. I., Semenova, T. P., . . . Miasoedov, N. F. (1998, January-February) [The anxiolytic action of an analog of the endogenous peptide tuftsin on inbred mice with different phenotypes of the emotional stress reaction.][Article in Russian] *Zh Vyssh Nerv Deiat Im I P Pavlova 48*(1):153–60.

Dihexa

Akhter, H., Huang, W.T., van Groen, T., Kuo, H.C., Miyata, T., & Liu, R.M. (2018). A Small Molecule Inhibitor of Plasminogen Activator Inhibitor-1 Reduces Brain Amyloid-β Load and Improves Memory in an Animal Model of Alzheimer's Disease. *Journal of Alzheimer's Disease 64*(2):447–457. doi:10.3233/JAD-180241

Benoist, C.C., Kawas, L. H., Zhu, M., Tyson, K. A., Stillmaker, L., . . . Harding, J. W. (2014, November). The procognitive and synaptogenic effects of angiotensin IV-derived peptides are dependent on activation of the hepatocyte growth factor/c-met system. *Journal of Pharmacology and Experimental Therapeutics 351*(2): 390–402. doi:10.1124/jpet.114.218735

Ho, J. K., & Nation, D. A. (2018). Cognitive benefits of angiotensin IV and angiotensin-(1–7): A systematic review of experimental studies. *Neuroscience & Biobehavioral Reviews 92*, 209–225. doi:10.1016/j.neubiorev.2018.05.005

McCoy, A. T., Benoist, C. C., Wright, J. W. Kawas, L. H., Bule-Ghogare, J. M., . . . Harding, J. W. (2013, January) Evaluation of metabolically stabilized angiotensin IV analogs as procognitive/antidementia agents. *Journal of Pharmacology and Experimental Therapeutics 344*(1): 141–154. doi:10.1124/jpet.112.199497

Wright, J. W., & Harding, J. W. (2015). The Brain Hepatocyte Growth Factor/ c-Met Receptor System: A New Target for the Treatment of Alzheimer's

Disease. *Journal of Alzheimer's Disease 45*(4):985–1000. doi:10.3233 /JAD-142814

Wright, J. W., Kawas, L.H., & Harding, J. W. (2015, February). The development of small molecule angiotensin IV analogs to treat Alzheimer's and Parkinson's diseases. *Progress in Neurobiology 125*: 26–46. doi:10.1016/j.pneurobio.2014.11.004

FGL-L

Asua, D., Bougamra, G., Calleja-Felipe, M., Morales, M., & Knafo, S. (2018, February 1). Peptides acting as cognitive enhancers. *Neuroscience, 370*, 81–87. doi:10.1016/j.neuroscience.2017.10.002

Barco, A., & Knafo, S. (2018, February). Editorial on the Special Issue: Molecules and Cognition. *Neuroscience 370*, 1–3. doi:10.1016/j.neuroscience.2017.11.003

Cognitive boost to brain connections. (2012 February 29). *Nature 483*(7387), 9. doi:10.1038/483009c

Knafo, S., Venero, C., Sánchez-Puelles C., Pereda-Peréz, I., Franco A., . . . Esteban, J. (2012, February 21). Facilitation of AMPA Receptor Synaptic Delivery as a Molecular Mechanism for Cognitive Enhancement. *PLoS Biology 10*(2): e1001262. doi:10.1371/journal.pbio.1001262

INDEX

TA1 (thymosin alpha 1)
    characteristics of, 84–89
    and TB4, 83
TB4 (thymosin beta 4)
    characteristics of, 81, 90–94
    and TA1, 83
TBI (traumatic brain injuries), 71
TCA cycle, improving, 26
tesamorelin, 51, 56–57
TFAM (transcription factor), 27–28
TGF-beta1, increasing, 44
TH2-dominance, 15
thalidomide, 8
therapeutics, peptides as, 9
therapy-induced senescence, 18
Thymalfasin (Zadaxin), 89
thymosin hormone, relationship
    to TB4, 90
TP-7, 112
tranquilizers, replacing, 112
transient senescent cells, 16–19
treatment protocols, side effects, 8

TReg cells, signaling to, 41
tuftsin, 112
type 2 diabetes
    cause of, 21
    and obesity, 107

U

ulcers of the stomach, preventing, 71
U.S. Food and Drug Administration, 7

V

viral pandemics, 39
virus, reaction of body to, 41

W

Western medical model, 3
wound healing
    accelerating, 71
    activating, 74

Z

Zadaxin (Thymalfasin), 89

ABOUT THE AUTHOR

William A. Seeds, MD is a board-certified surgeon practicing medicine for over 30 years. He is Founder and Chairman of the International Peptide Society, Faculty Developer and Lecturer of the A4M Peptide Certification Program, and a leading peptide therapy researcher. He is Chief of Surgery and Orthopedic Residency Site Director for University Hospital, Conneaut, and Medical Director of Orthopedic Rehab and Sports Medicine at the Spire Institute, a USA Olympic training site. Dr. Seeds has been honored at the NFL Hall of Fame for his medical expertise and in treating professional athletes, and serves as Professional Medical Consultant for the NHL, MBL, NBA, and NBC's Dancing with The Stars.

Printed in the USA
CPSIA information can be obtained
at www.ICGtesting.com
LVHW072334170324
774729LV00014B/217